Rethinking Risk Management

Rethinking Risk Management

Critically Examining Old Ideas and New Concepts

Rick Nason

BUSINESS EXPERT PRESS

Rethinking Risk Management: Critically Examining Old Ideas and New Concepts

Copyright © Business Expert Press, LLC, 2017.

First published in 2017 by
Business Expert Press, LLC
222 East 46th Street, New York, NY 10017
www.businessexpertpress.com

ISBN-13: 978-1-63157-541-9 (paperback)
ISBN-13: 978-1-63157-542-6 (e-book)

Business Expert Press Finance and Financial Management Collection

Collection ISSN: 2331-0049 (print)
Collection ISSN: 2331-0057 (electronic)

Cover and interior design by Exeter Premedia Services Private Ltd., Chennai, India

First edition: 2017

10 9 8 7 6 5 4 3 2 1

Printed in the United States of America.

Abstract

Risk management has become a key factor of successful organizations. Despite risk management's importance, outdated and inappropriate ideas about how to manage risk dominate. This book challenges existing paradigms of risk management and provides readers with new concepts and tools for the current dynamic risk management environment.

This book has two major origins: The first is a series of executive workshops that I have been conducting for the last several years for a major international company. The second origin is an innovative and popular course on enterprise risk management that I have developed and delivered for MBA students. The book reflects these two origins in that it covers both the current base of risk management knowledge but critically examines that base by exploring emerging risk management ideas and concepts. The framework for the book is a series of questions that allows for an interesting and thought-provoking look at current ideas and forward-looking concepts.

This book, intended for senior managers, directors, risk managers, students of risk management, and all others who need to be concerned about risk management and strategy, provides a solid base for not only understanding current best practice in risk management, but also the conceptual tools for exploiting emerging risk management technologies, metrics, regulations, and ideas. The central thesis is that risk management is a value-adding activity that all types of organizations, public, private as well as not-for-profit, can use for competitive advantage and maximum effectiveness.

Keywords

ambiguity, complexity, decision making, enterprise risk management, hedging, management of complexity, regulation, risk management, risk mitigation, strategic analysis, uncertainty, volatility, VUCA

Contents

Acknowledgments

This book arose largely out of the many interactions I have had with a variety of risk management clients. To them I am very grateful for all of the ideas and discussions. Stephen McPhie, my partner and cofounder of RSD Solutions, has been an excellent source of ideas and a great help in clarifying my ideas and thoughts on risk management.

John Fraser, the retired Chief Risk Officer of Hydro One, has been a terrific risk educator for me. There is no one else I have met who brings the same amount of passion and wisdom to enterprise risk management than John. Although we do not always agree, I have learned more about risk management from discussions with John than I ever did through more formal study.

Another source of inspiration for me has been the many graduate students and executive education students who have taken risk management courses with me. They never fail to challenge me and force me to look anew at my ideas.

Finally, but certainly not least, I would like to thank my family for permitting me the time to complete the work necessary for this book and for allowing me to test my ideas on them. They are the ultimate source of inspiration for everything I do.

Introduction

Another Book on Risk Management!

Did you just groan at the thought of opening another book on risk management? Like election cycles, media talk about celebrity hairdos and cute cat videos trending on social media; perhaps, the least interesting thing might be another book on risk management. However, you got past the first two sentences, so possibly you believe there is something still needed in risk management thinking that is keeping you going.

Risk management has always been a part of business, as it should be. However, it appears that the limelight on risk management seems to go in cycles—and always at the most inappropriate times. By this, I mean that the attention on risk management always occurs after an assumed failure of risk management. In other words, the attention on risk management is always ramped up after there has been some sort of crisis. A financial crisis occurs—risk management (and associated regulation) comes into the news cycle as soon as the dust has settled. A natural disaster occurs— risk management (and associated regulation) comes into the news cycle as soon as the dust has settled. A terrorist attack at a public facility—risk management (and associated regulation) comes into the news cycle as soon as the dust has settled. An industrial accident—well you get the picture.

The unfortunate aspect of risk management is that it predictably comes to the fore after some form of bad thing has happened. In some sense, risk management has kind of become the undertaker of business. If a death has occurred, the undertaker appears; if a crisis has appeared, the risk manager appears, (and their sidekick a regulator with a new process to prevent a reoccurrence). Although surveys show that dentists tend to be the professionals who commit suicide with the greatest frequency, I suspect that risk managers must not be too far behind. After all, who wants to be known only as the person who both cleans up after a mess *and* takes the blame for allowing the mess to occur in the first place? Not a cheery perspective to take.

To me, this seems the wrong way to think about risk management, and thus, the reason for writing this book. I believe that risk management is a very valuable and necessary task that every business needs to focus on in some way, but the focus has, for the most part, been all wrong, and frequently counterproductive, or at least inefficient.

In all value-creating tasks, there is a need to periodically revaluate assumptions—those that are both implicit as well as explicit—and ask questions about whether there is a better way forward, or perhaps even a paradigm shifting way to do things differently and better. I believe that this is a good time to do so for risk management. Risk management (and its associated regulation) is choking business and nonprofits with little discernible benefit. Few companies or organizations are calculating the cost to benefit analysis, both because they do not know how to do the calculation, and because they subconsciously realize that they do not want to know the answer. Perhaps more worryingly, most companies and risk managers are following "best risk practices" without ever questioning why. The risk management profession in many instances is making the mythical conforming lemmings look like a herd of feral cats. (Perhaps, the phrase "best practice" is the most evil phrase in business or at least in risk management.)

This book is centered on a series of questions whose purpose is to re-examine some of the underlying assumptions behind much of modern risk management. Furthermore, these questions form the basis of examining other ways forward for risk management. Even if one accepts the dubious assumption that risk management is not broken, there will still be value gained from such an examination, if it sparks ideas for improvement.

Chapter 1 begins with a very simple, very fundamental, but also a very critical question—what is risk? Perhaps, the question is trivial, but in workshops I have conducted, I get a wide variety of responses, and often, a passionate debate will break out among the participants. Apparently, the answer to the simple question of "what is risk?" is not so simple. Of course, this leads to a follow-up question of what is risk management? Another simple question that proves to have a not-so-simple answer—or at least not simple when you hear managers try to explain what risk management is. The physicist Richard Feynman supposedly once claimed that if you cannot explain what you do to someone in grade six, then you do not know what you are talking about. How much of risk management can be explained to a preteenager in grade six much less the managers or even the board of directors of an organization?

It is often only by going back to first principles, or a clean sheet of paper, that one can again make things simple again, and provide the clarity that is needed to develop more productive and effective paths forward. Asking two simple questions of what risk is and what risk management is helps to start us off on that process.

How often does someone in your organization question the reasoning behind a certain process? I suspect it is quite often, but they keep their questions to themselves without taking the risk of openly questioning and potentially exposing their lack of knowledge. Perhaps, you are one of those doing the silent questioning. Generally, if someone actually does verbalize the question, rather than leaving it unspoken, there is frequently no response other than "We've always done it this way." Not a very satisfying response! Progress is not based on always and unquestioningly doing things the way they have always been done. Sometimes, processes are based on assumptions that turn out to be false upon close examination. However, without questioning these assumptions or axioms, the situation will never improve, unless by some fortuitous accident, which is not a high-probability strategy. It becomes very easy to lose the plot of what the main objective is without an occasional pause to ask what the end goal is, and what the best path to get there is. Chapter 2 openly questions whether or not we have lost the plot for risk management and the role of making assumptions about false risk axioms has in this.

Chapter 3 starts a discussion about the emerging (no pun intended) science of complexity. In common language we often commingle the words complex and complicated, but to a systems scientist, the two words and their associated systems have very different meanings. Along with a difference in meaning is a difference in how each comes about. Furthermore, there are dramatic and often counterintuitive methods for dealing with each. Things that are complex are most definitely not complicated, and more so, they must be managed in a very different manner. I believe that complicated thinking, when complexity thinking is called for, is the biggest mistake, the most common mistake, and the most serious flaw in most of risk management as it is currently practiced. The problem is that almost all conventional responses in risk management are based on flawed complicated thinking. Chapter 3 challenges this line of thinking and presents a radically different way of thinking about most risk management situations.

I am assuming that seeing as you were intelligent and conscientious enough to buy this book that you are never a source of risk in your organization or even in your personal life. You are not the source of risk, but something is—and what is that something? In other words, where does risk come from? Is risk just random, does risk come from the Gods, does risk only occur when processes are not followed? If you do not know where something comes from, how are you going to stop it or change its path or even exploit it? Chapter 4 has the nerve to ask the question of where risk comes from. The answer to the question (which of course will be different for different organizations) then guides the development of an appropriate response to risk, rather than developing a response and hoping it is suitable for the issue at hand. (By the way, hope is neither a very prudent nor a very mature risk strategy.)

Are risk frameworks evil? Chapter 5 critically examines the role that risk frameworks play in preventing risk management from achieving its potential. Risk frameworks seem to be the rage among risk consultants, those who engage risk consultants, and those consultants who are actively engaged in risk management societies. However, are risk frameworks useful for anyone besides consultants? Frameworks definitely have their place, but that place needs to be well defined and constrained. Utilizing a framework without thinking about the consequences—both positive and negative—is simply not smart risk management.

Is your risk management function a cost center or a profit center? Does your risk management function add value to your organization? If so, how much value? Is risk management part of the strategy or a series of processes to ensure that the strategy stays on track? Do risk managers get compensated on value added or on the prevention of losses? After covering the various definitions of risk presented in Chapter 1, it will be seen that the answers to this question come directly from how the risk management function is viewed. In most organizations that I work with, risk management is most frequently seen as the "Department of No!"; the high-profile function with the support of the board that puts the brakes on good ideas. Changing the focus to a value-creation center; the "Center of How to Do Things Better" makes a huge difference. How to bring about this change is the topic of Chapter 6—Does Risk Management Add Value.

Chapter 7 starts with the story of Tomas Lopez, a young lifeguard who had the audacity to willfully flaunt and ignore the rules of his position and actually save someone who was swimming outside of the designated swimming area. The question for Chapter 7 is whether risk management should be process-based or judgment-based. As risk management becomes more of a specialty, the prevalent thinking is that it is too complicated for the unwashed nonspecialist to manage, and thus, it must be made as idiot-proof as possible; thus, the rise of regulations and processes and the demise of allowing the people who actually do the work to use judgment. While processes certainly have their role, just like a Shakespearean actor, their role has a specific place and a specific time to appear on stage.

Risk management, along with every other important organizational function, does not exist in isolation. All the functions of an organization exist within the context of an organizational culture. Chapter 8 examines the question of how one goes about creating an effective risk culture. Culture is something like the weather; something that a lot of people spend a lot of time talking about, but alas, something that they seem to be helpless to affect. While I do not know if there are any magic steps to creating an optimal culture, there are certainly some steps to take that help avoid a poisonous environment.

Risk homeostasis is not a term that rolls off the tongue of many risk managers (nor their associated regulators), but it is an important phenomenon that keeps rearing its insidious head time and time again. Risk homeostasis is the answer to the question of Chapter 9—Can Your Risk System Be Too Good. In a nutshell, risk homeostasis is when risk management actually creates more risk due to the presence of a strong risk management infrastructure. In essence, in the act of being so careful about avoiding risk, you actually create a set of riskier outcomes. A bit of irony that is not at all funny.

This book concludes with a look at some future scenarios for risk management. Predicting the future is fun to do, although not all that useful as Orgel's law teaches us that "evolution is smarter than we are." Creatively thinking about the future does, however, allow us to profit from what I like to think of as the first law of risk management; the mere fact that you acknowledge that a risk exists allows one to automatically increasing the probability and the magnitude of it occurring if it is a good risk while simultaneously decreasing the probability and severity of it if it is a bad risk.

As I write this, I am listening to a news report of an assumed shooting with LAX airport. The reports of a shooter fortunately turned out to be false, but that did not mean that there wasn't a scene of chaos as passengers ran for their lives and authorities tried to clear the airport. What was interesting, however, was one of the many terrified passengers who were interviewed by the media. One such passenger was asked, "Did you hear gunshots?," and his response was "No?" The natural follow-up question that was asked was "Then, why were you running?," and the response was "Because everyone else was!" Now I am not suggesting that if you see a pack of people running for their lives that you should refrain from following suit, but it appears that a mass panic ensued for no apparent reason. In a similar fashion, that seems to be the situation in risk management as well. We calculate the same risk metrics and employ the same risk tactics as everyone else, solely in large part because that is what everyone else is doing.

The follow-the-crowd strategy for risk management has several obvious flaws. The first is that one-size-for-all risk management is not likely to be optimal. Different companies have different types of operations and different tolerances and preferences for risk. They also have different abilities to deal with and manage risk. It is like going into a clothing store and simply buying the most popular outfit in the most popular size. While this might be a suitable tactic to allow one to "fit-in" when they are in junior high school, it is obviously a foolish and childish thing to do once one has reached a certain age of maturity and emotional intelligence.

A second and more serious issue with the follow-the-crowd strategy is that it is a brain-dead tactic that does not encourage thinking, learning, or intelligent analysis. How is one to discover new paths if one simply follows the crowd? How is one to learn and develop additional skills if letting the crowd carry you away is the main operational method?

Into this atmosphere, the academic and the training industry are producing more highly and rigorously trained risk managers than ever before. The number of professionals with a risk management certification has risen significantly over the last decade, while the number of academic programs with risk management in their title is also growing rapidly. Clearly, there is a demand (and a supply) for participating in some form of formal risk management training. Managers (and prospective managers) obviously want a roadmap so as to know how to follow the crowd.

The demand for risk management training raises an interesting question of whether risk managers can be developed. In other words, is risk management an innate skill or is risk management a well-defined process that can be trained?

With all of this attention on risk management, it behooves one to ask the question of organizational leaders if they believe that their organizations are getting better at risk management. To paraphrase Ronald Regan's election-winning phrase from 1980, "Are you better off than you were 10 years ago in terms of your risk management?" What would you answer for your organization? Hopefully, that was not too sobering of a thought.

If your current risk management paradigms are not doing it for you, what do you have to lose by trying something different? After all, there is the well-known and prudent rule of if you find yourself in a hole, the first thing you should do is to stop digging.

This book is based on three fundamental tenets: (1) that risk management is a vital task for developing competitive advantage, (2) having knowledge, but more importantly skill and intuition in risk management is key for advancing one's career, particularly in light on the onslaught of the robots and computers that are replacing both blue-collar and white-collar jobs, and (3) there is a need to take a fresh look at risk management by questioning some old assumed axioms and asking some fresh questions.

I hope you have as much fun in reading this book as I had writing it.

CHAPTER 1

What Is Risk?

What is risk? It is such a simple question. It is a question that anyone over the age of three can answer. The answer, however, is also the basis of most of the current problems and inefficiencies in risk management today. Humor me for a second and take a moment to say out loud what your definition of risk is. (Okay, if you are too risk-adverse to chance being heard talking to yourself, you can simply think it out loud in your head.)

When doing risk workshops, it is always quite telling how people will squirm and fidget in their seat as I ask them what their definition of risk is. Workshop participants then go to great lengths to avoid eye contact, so that they will not be selected to give their definition of risk. Again, this is a definition that almost every three-year old can articulate. It then starts to get really interesting as workshop participants strive to demonstrate their risk prowess by seeing who can come up with the most precise, and consequently, the most academic definition. Inevitably, the answers start to get more quantitatively oriented, or more regulatory and legalistic in nature. Carry the conversation on long enough and you will need a doctorate in math and or laws in order to make sense of it.

You might think that it is absolutely bizarre, or at least a waste of space by spending a chapter talking about the definition of risk. However, I believe that having a clear and consistently understood definition of risk throughout an organization is one of the easiest, yet one of the most beneficial steps that a company can take in improving their risk management activities. Conversely, not having the right definition of risk, and not having a clear and consistently understood definition is the root cause of many of the problems in risk management.

If you ask the average person on the street what their definition of risk is, they will likely respond that risk is a chance that something bad will happen. Indeed, if you look up risk in the dictionary, you will get, "the possibility that something bad or unpleasant (such as an injury or a loss) will happen."[1] This definition is fine and good, but it leaves a lot to be desired for risk management purposes.

Firstly, the definition of "possibility of something bad happening" is not consistent with the mathematics of the most common ways to measure risk. Secondly, it is an extremely limiting definition that forces most of the potential value of a risk management function wasting away. Thirdly, and perhaps most importantly, it is a very negative definition, which, in turn, imparts a negative pall over all of risk management. Finally, this common (mistaken) definition of risk is one of the reasons I decided that this book needed to be written.

Rethinking the Definition of Risk

The Chinese symbol for risk is often cited as being composed of danger and opportunity. This is a much more enlightened definition, as well as a much more useful and productive definition for risk management purposes. Another way to state this is to define risk as the possibility that bad or good things may happen.

You might be thinking that defining risk as "the possibility that bad or good things may happen" is a convenient butchering of the English language. While it is true that the editor of this book will find many instances of my butchering of the English language, the definition of risk as given is perfectly legitimate and valid in the context of organizations and in the context of risk management. Indeed, as will be argued, it is, by far, the preferred definition. It is also the de facto mathematical definition of most risk management measurements—whether you realize it or not.[2]

---

[1] http://www.merriam-webster.com/dictionary/risk.

[2] Almost all quantitative risk measures are related to variance or standard deviation of outcomes, which implicitly assumes that outcomes can be positive or negative. Unless your organization (or regulator) uses measures based on semi-standard deviation, or semivariance, then you are implicitly measuring risk using the good and bad things may happen definition.

The Committee of Sponsoring Organizations of the Treadway Commission, more commonly known as COSO, is a joint initiative of a variety of organizations with a common interest in developing standards and frameworks for effective risk management. The COSO framework for enterprise risk management is considered by many to be the definitive framework for risk management. Their definition for risk is:

> *Enterprise risk management is a process, effected by an entity's board of directors, management and other personnel, applied in strategy setting and across the enterprise, designed to identify potential events that may affect the entity, and manage risk to be within its risk appetite, to provide reasonable assurance regarding the achievement of entity objectives.*[3]

A moment's reflection will show that it is consistent with the proposed definition of risk, while albeit being a bit more sophisticated in its wording.

There are three elements to my proposed definition of risk; firstly, there is an element of uncertainty, secondly there is an element of the future, and thirdly, and perhaps most significantly (and unconventionally), there is an element that risk has both upside and downside components.

The first element of the definition involves uncertainty. In simple terms, with risk, there is an element of not knowing what will happen. Frequently, risk and uncertainty are used as synonyms for each other. When we state that a situation is uncertain, we may as well say that it is a risky situation and viceversa.

In the technical mathematical literature, as well as in the academic risk community, risk and uncertainty have close, but different meanings. With risk, you have a range of possible outcomes, but the mathematical distribution is known. For instance, we can state what the average daily returns of the S&P 500 index were, as well as what the standard deviation for the index was, for any given year for which we have the

---

[3] http://www.coso.org/documents/coso_erm_executivesummary.pdf.

data.[4] However, what will be the most popular genre of music among high school students 10 years hence is, however, unknown, and at the present time, unknowable. It is quite possible that the most popular genre of music 10 years hence has not yet been conceived of. This is uncertainty. Given the pace of technological change in the music industry, it is also hard to envision how that music will be distributed and consumed as well. Thus, we see that stock returns can be labeled as risky, while future popular music genres are uncertain.

This parsing of the difference between risk and uncertainty is, for the most part, academic and does not really need to concern us. There is, however, one catch: risk management as a profession is becoming enthralled with what can be quantified, and thus, there is an inherent biases toward elements that can be measured with precision, which includes events for which a known mathematical distribution such as the normal distribution or a Poisson distribution (or perhaps even something sexier sounding such as a Gaussian distribution) can be applied to. Experienced organizational professionals will immediately see a problem with this, as many key risks cannot be so easily distilled to such a known distribution. As management guru Peter Drucker famously stated, "what gets measured gets managed." However, that does not mean that all that needs to be managed gets measured, or conversely, that everything that can be measured needs to be managed. It might be argued that the most important risks are precisely those that cannot be measured—a theme that we will return to later.

The second element of our definition of risk is that it concerns the future. That may be an incredibly obvious statement, but too often we are focused on the past; the past crisis, the past mistake, the past regret, and as a result, spend way too little time creatively thinking about what might be. Risk cannot be managed in the rearview mirror, and while the past might be the best precursor to the future, it is not a very reliable one. The

---

[4] I am playing fast and loose with the facts here. Later, in Chapter 2, it will be discussed that in reality, stock returns are not normally distributed (as is commonly assumed and is implicitly implied by using the standard deviation as a useful parameter), but, in fact, are sneakily leptokurtic, but that is a subject for Chapter 2.

professional risk manager must be knowledgeable and respectful of the past, but with a vision focused on the present and an imagination trained on the future. Again, there seems to be a disconnect as risk measures and risk strategies, as well as tactics are frequently (and correctly) criticized for fighting the last crisis. We cannot change the past. The best we can hope for is to manage the future, but obsessively focusing on the known past is not only too convenient and intellectually lazy, but it takes valuable time and energy away from imagining future scenarios that will need to be managed.

The final element of our definition of risk is that risk involves both the downside as well as the upside. Risk is the possibility of both good things as well as bad things happening. This is the element of the definition that causes pause with the most people. In workshops, it is the element that causes the most push-back, as many risk managers, as well as regulators, believe that the sole function of risk management is preventing the downside. It is certainly how almost all risk management functions in practice. However, an equal focus (if not more than equal to overcome long standing biases) on the upside is what makes all the difference.

Why the Definition Is Important?

So why does the definition matter? It matters because of attitude, execution, and effectiveness.

Let's start with attitude. My experience has been when talking with frontline professionals that they in some way label the risk management function of their organization as the "Department of No!" Admit it, that is likely how the risk management department in your organization is thought of. Not a very optimistic thought, is it? It is kind of like risk management being the dark cloud; the spoil sport of the group; the "it's sure to rain" declarer. No one wants to invite the party pooper to the party.

A change in the definition is more than a slogan. To begin with, a change in definition can become an attitude changer. In Chapter 8, some ideas for creating a good risk culture will be discussed. This simple, but profound change in definition is a key element in creating a positive and effective risk culture. Instead of a culture of fear, or blame, or a culture of restraint and being held back, the risk culture becomes focused on the positive, the possibilities, and on how risk adds value and effectiveness to the goals and objectives of the organization. This does not diminish the focus on the downside, but counterintuitively can enhance an appreciation for, and the understanding of managing both the downside and the upside both individually and in tandem.

Secondly, a change in the definition helps risk to become a proactive function, rather than a reactive function. Think for a second about the tasks that bring out the best of your procrastination skills. Do you procrastinate on the positive things, the optimistic things, or do you procrastinate more on the perceived negative or downside events? Do you procrastinate more on making your dental appointment or booking tickets to see your favorite sports team play in the finals of the championship? There are upsides and downsides to both events. My dentist recently told me that my teeth are as solid as rocks, and your dentist as well may give your mouth a clean bill of health, but instead you focus unwarrantedly on the need for a potential root canal. Meanwhile, your team may suffer a blow-out in the playoffs, but instead you focus on the joy and thrill of victory and the experience of celebrating with a group of like-minded fans. You do not allow the thought of your team suffering an embarrassing beat-down to enter into your consciousness. It is easy to see why the dentist has a receptionist phone your office to coerce you into making your semiannual appointment while there will be a queue of people camping out overnight in order to buy playoff tickets.

A large part of risk management is dealing with human nature. Having a definition that incorporates both the positive and negative elements of risk works with human nature to produce a far more proactive attitude toward risk. It is human nature to focus on fear and downside risk, unless a more positive element is also explicitly introduced.

A central theme of this book is that risk management function should be a value-oriented function, rather than a cost center. Have some patience and before you start espousing all those studies about companies losing their figurative shorts by making the risk function a profit center, humor me for a few chapters, so that I can explain and build the argument. I am not advocating that risk managers should start trading derivatives in order to time the markets and make exceptional profits. As someone who is trained as a finance professor, and as someone with professional trading experience, I have a strong belief in the efficient markets hypothesis, which states that it is impossible to make positive abnormal returns from financial trading. The exploits of firms such as Procter and Gamble and Metallgesellschaft[5] in the 1990s have unambiguously and definitively shown the folly of corporate entities trying to make money solely through sophisticated financial trading, rather than efficiently making things and selling things. What I am advocating is using intelligent and positively focused risk management to enhance the effectiveness and profitability of making things and selling things and services. More on that later, but the point for now is that risk management should be seen as a value-creation activity. Instead of the "Department of No!," risk management has the capability to become the "Department of How We Can Do It Better!" Many of the techniques and tactics for managing downside risk can also be, and should be, applied toward enhancing upside risk. The proposed definition of risk goes a long way toward allowing this to happen.

The final advantage of the proposed definition is that it changes the definition of what the "risk management" function is. Defining risk as the possibility of bad things happening sets up risk management as the function to prevent losses. Setting the definition of risk as the possibility that bad or good things may happen provides the basis for a much more positive and valuable objective for risk management.

The Definition of Risk Management

If risk is the possibility that bad or good things may happen, then risk management becomes managing so as to increase the possibility and magnitude of good risk events happening while simultaneously managing so as to decrease the possibility and severity of bad risk events happening. The change is simple, subtle, but critically important in dramatically increasing the value and effectiveness of the risk management function in any organization.

---

[5] Procter and Gamble lost significant sums with an interest rate hedging program that was designed to save them a few basis points on their borrowing costs. Metallgesellschaft likewise lost significant sums with an aggressive strategy of hedging oil.

As an example (albeit a trivial one), consider the last time that you took a trip in your car. It could have been a cross-country trip with your family or it could have been a five-minute drive to the grocery store. Assuming that you are an experienced driver, and a good driver,[6] you almost certainly practiced risk management as I have proposed. You drive defensively to prevent an accident, but you also drive so as to achieve your objective, namely arriving at your destination in a time-efficient manner. If you were acting like most risk management departments, you would take far fewer trips to avoid the chance of an accident. No car trip means no car accident, but also no family viewings of the Grand Canyon or no groceries at home.

Often, when I propose this definition of risk, it will be argued that it is axiomatically impossible to simultaneously manage to increase upside probability of good events while decreasing the possibly of downside events. That argument of course is poppycock and a sign of lazy thinking. We are constantly trying to increase our odds of success while simultaneously decreasing our odds of failure. We do this with almost every activity we undertake. In fact, by focusing on increasing success, we automatically are decreasing the possibility of failure. In school, the best way to avoid failing a course is to work to get a great mark in the course. At work, the best way to avoid getting fired (or demoted) is to work so as to get promoted. In sports, it is often stated the best defense is to have a great offense. Winning is not just preventing your opposition from scoring, it also means scoring yourself. However, how focused is your risk management function on scoring?

If you say that scoring (i.e., creating profits) is the function of operations and marketing, then you are missing the point. In saying that organizations are siloing risk management, while probability also spouting a nice platitude about how risk management is everyone's responsibility.

---

[6] Interestingly, when asked about their driving skills, a large majority of drivers respond that they are well above average when it comes to driving skill and with being a safe driver. Humorist Garrison Keillor called this the Lake Wobegon effect, where school districts report that almost all of their students are above average. Of course, this is a mathematical impossibility, but an important risk management effect caused by overconfidence.

I believe that risk management truly is everyone's responsibility, but having a risk function that focuses solely on the downside almost always produces the opposite effect. Assuming that risk management will take care of the bad stuff and marketing will produce the profits is not an effective integration of functions. In Chapter 9, we discuss one unintended and counterintuitive consequence of this, which is known as risk homeostasis: namely having a really strong risk function focused on the downside actually increases the probability and severity of something bad happening.

Defining risk management as increasing the probability and magnitude of good risk while decreasing the probability and severity of bad risk implies balance, and risk management is nothing if not an exercise in balance. It is a balance between art and science, process and judgment, knowledge and intuition, people and processes, and the current and the future. It is extremely difficult to have balance when one is so unbalanced by focusing solely on the downside.

With the balance of considering both upside and downside risk, the tactics and techniques of risk management can be applied for the upside as well. To take a simple example, the techniques of risk measurement can be used to measure upside. Take, for example, the often-used measure of value at risk or VAR, which is primary metric used to measure downside risk in financial institutions. An informal survey of institutions shows that almost no one uses VAR to measure the upside potential of proposed transactions or strategies. However, firms have a battalion of quantitative analysts refining downside VAR. A simple and straightforward application of downside VAR to upside VAR can yield surprising and valuable metrics for assessing and implementing proposed strategies. While a few institutional investors look at upside VAR versus downside VAR, the number doing so is few. This is just one of many different risk techniques for which the potential applications and tools of risk management are literally only being half utilized. More examples will be provided later in this book, but the point here is to quickly point out here the folly of focusing solely on the downside.

Finally (for now), rethinking the definition of risk management allows for a more enhanced set of responses to risk. Traditionally, risk management actions are limited to eliminate, avoid, outsource, or mitigate. Rarely is the embracing of risk considered an appropriate response to risk. However, the examples are numerous, where the potential for upside should be embraced.

In his popular and insightful book "Antifragile: Things That Gain From Disorder,"[7] author and risk management guru Nassim Taleb points out the need to think about instances where risk should be embraced and not avoided or mitigated. While there are many more ideas in Taleb's book, the central thesis is that there are many instances in which risk or volatility are good things and that the inherent risk and volatility should be exploited, and not quashed. One example from the world of sports are those athletes and teams who play so as to not lose, versus those that play to win. While both teams face the same amount of risk and uncertainty, it is a common axiom in sports that those who focus on winning and embrace the possibility of winning become more successful than those who focus on losing, and thus play more conservatively. In risk management, as in sports, embracing the upside can trump protecting against the downside.

The Focus Gets Managed

What you focus on is what you see and what you will manage. I will talk at length later about the focus on measurable risk, but let's start with a simple experiment. On your way to work today, how many red Volkswagens did you see? You cannot answer the question, can you? Now assume that I rush into your office five minutes before you leave for the day and I shout at the top of my lungs five times the phrase "red Volkswagen." I bet that you will spot numerous red Volkswagens over the next week. My shouting "red Volkswagen" did not change the number of red Volkswagens on your path, it simply changed your focus. Take another simple example. Think of the last time you bought a new car. You were probably proud of the fact that you were unique, stylish, and one of a very few number of people who had the means and the good taste to buy such a classy car. However, over the next few days after buying your new car, it

---

[7] Nassim Taleb, "Antifragile: Things That Gain From Disorder," Random House, 2012.

seems that all you see is cars that are identical to yours. You do not feel so unique and special anymore, do you?

This very simple example shows the power of focus. I used to do the red Volkswagen experiment at workshops until participants started to get annoyed that all they could spot thereafter was red Volkswagens. (I still run into former workshop participants that tell me they can still spot red Volkswagens from a mile away.) However, the point is that if risk management focuses on bad risk, then all it will see is bad risk, while missing the more valuable good risk.

Psychologist Richard Wiseman performed a very interesting series of experiments on luck.[8] With a group of research subjects, Dr. Wiseman asked them whether they thought they were generally a lucky person or an unlucky person. You might think that is a very arbitrary, and perhaps irrelevant characterization, but his results are fascinating. Dr. Wiseman put the subjects through a set of exercises. In one exercise, he promised them a monetary reward if they could correctly count the number of times a symbol appeared in a specific section of a newspaper. On the second page of the newspaper section he gave them, he had printed a full page advertisement that told them the number of times the symbol appeared and also promised them double the reward if they stopped counting at that point. A large percentage of the "lucky" people stopped at that point and collected their double reward, while virtually all of the unlucky people continued to examine the rest of the newspaper. A powerful illustration of keeping an open mindset and perhaps an explanation for why so many organizations known for their impressive risk management departments get involved in risk debacles. More on this in Chapter 9, which hopefully incents you to keep reading.

Risk is dynamic, not static; seems obvious, and it should seem obvious. However, constraining risk to the downside takes away the dynamism. With the focus on the downside, there is a focus on preventing recurring events—preventing the downside risk that happened in the past. It is the main reason behind risk managers being criticized for managing the past crisis, not the forthcoming issues.

---

[8] Richard Wiseman, "The Luck Factor: The Scientific Study of the Lucky Mind," Cornerstone Digital (Kindle Edition), 2011.

Concluding Remarks

William James, considered by many to be the founder of the field of psychology, wrote that "pessimism leads to weakness, optimism to power." Focusing strictly on the downside is limiting, while focusing on both the upside and the downside of risk is empowering. It expands the possibilities for risk management and makes risk management a much more powerful part of an organization's success.

Admittedly, it is true that often there is more organizational risk on the downside than on the upside. For instance, in credit risk, there is little upside risk; there has never been a case of a bank borrower paying back more than the required principal and interest of their loan. However, whenever and wherever appropriate, risk and risk management should be thought of as two-sided. Redefining risk as the possibility that bad or good things may happen, and redefining risk management as a function to focus on the upside as well as the downside, is a game changer and a necessary paradigm shift for effective risk management.

CHAPTER 2

Have We Lost the Plot?

An axiom is a statement or a fact that is universally accepted to be true. Axioms are a concept that originated in mathematics and logic. For example, there is a reflexive axiom that says that "a" equals "a". In other words, things that are equal to the same thing must be equal to each other. It is so trivial that it is actually difficult to state in words. But, we all know and understand without questioning it that five equals five.

Axioms prove to be extremely useful in developing mathematics as a field of study. All of the math that you learned in school is based on a few basic axioms. In fact, if you took a basic math theory course in high school or university you probably recall memorizing Euclid's axioms and postulates. Without having to reprove these axioms each time, the development of extremely complicated concepts can be developed and expanded upon very efficiently.

In life, we also have axioms; the sun will rise in the East, apple pie is good, and the cable repairman will be late for their scheduled appointment, and you will waste a good portion of your day waiting for them to show up and implement a five minute fix. As in mathematics, axioms in life tend to be useful, although in life, we are not often trying to be as rigorously precise as in mathematics. Axioms, however, are only useful to the extent that they are true. Acting on a false axiom can lead to mistakes, incorrect conclusions, mishaps, and unintended consequences.

Risk management also has a series of axioms as well. However, with risk management, we tend to be quick to assume a hunch is truly a truth or something that we should accept as an axiom. Also, we are quick to build a series of more complicated theories and strategies for risk management upon these unexamined, and perhaps shaky, half-truths that we all too readily accept as axioms.

The purpose of this chapter is to more critically examine some of these axioms and the possible effects and unintended consequences that may arise from so readily accepting these false axioms. It seems at times as if we have lost the plot in managing risk. We have become a slave to perfecting the axioms rather than striving for better ideas for managing risk. Let's begin to explore some of these false axioms.

Frequentist Statistics

Let's play a little gambling game. I will flip a coin in the air 1,000 times. Each time it comes up heads, I will pay you $1 million. However, each time it comes up tails, you will have to pay me $900,000. We will settle up the net amount won and lost on each flip at the end of the 1,000 flips.

Would you enter into this game of chance with me? Of course, you will—or at least you should! On each flip, you expect to win on average $50,000. Yes, there will be times that you lose on the flip, but if we flip the coin 1,000 times, you expect that the coin will come up heads approximately 500 times and tails 500 times. Calculating your expected net winnings and your net losses you calculate expected net winnings over the 1,000 flips to be $50 million. You will be set on easy street for life and I will need to find some deep pocketed financial backers. If you are really mathematically inclined, you might calculate the value at risk for your gambling position, but assuming a fair coin (am I perhaps trying to deceive you with some weighted coins?), you will quickly calculate that the probability of your losing money is exceedingly low. You will take me as a sucker and plan your early retirement and life of future luxury.

Let's however change the game ever so slightly. It is boring and tiring to flip a coin 1,000 times, so let's just flip it once and multiply the outcome by 1,000. Mathematically, the expected outcome is the same, so it should not be a problem for you to readily accept this slight change. In fact, you will get your expected winnings that much quicker, as you do not need to wait for all of that flipping to be conducted.

Despite your state of excitement about becoming instantly rich, you have a nagging feeling about agreeing to the gamble, as you should. With the one-time flip, although the expected outcome is the same, the odds of your losing money are now 50 percent. Furthermore, the amount of money you will lose with a 50 percent probability is enough to bankrupt you unless you are one of the approximately 2,000 billionaires who exist in the world.

What is the difference in the two gambles? The difference is frequentist statistics. As brilliantly explained by finance professor and author Riccardo Rebonato in his enlightening book *Plight of the Fortune Tellers*,[1] frequentist statistics are frequently abused in risk management. Frequentist statistics is when we assume that an event is repeated multiple times. In such cases, and assuming we know what the underlying statistical distribution of outcomes is (a sometimes very bold and incorrect assumption), we can use statistical analysis to determine what the outcomes are likely to be.

Frequentist statistics, for instance, is the basis of actuarial science. For any given person, an actuary cannot tell whether the person will die in the next year or not. However, given a large number of people with similar characteristics, say a collection of 1,000 men between the age of 60 and 65, in a given socioeconomic status and in otherwise good health, the actuary can predict quite accurately how many of the 1,000 men will die within the next year or even the next five years. This of course is how, in part, life insurance rates are calculated. Frequentist statistics is also the basis behind using credit scores for credit analysis for the granting of credit cards and other type of risk events involving large numbers of transactions or interactions.

Frequentist statistics is also the basis for much of risk management analysis. Virtually, every program or course in risk management begins with a session on statistical analysis. In fact, as I write this chapter, I am taking breaks from preparing lectures for a Masters level university course in risk management on statistical analysis. The use of frequentist statistics in risk analysis is so ubiquitous that it is axiomatic that the use of statistics is a valid and valuable risk management tool. However, this is an axiom that needs to be questioned.

There are four critical problems with frequentist statistics. The first is that in life, and particularly so with major decisions that have long-lasting effects, we rarely, if ever, get to make the decision multiple times. In the

---

[1] Riccardo Rebonato, "Plight of the Fortune Tellers: Why We Need to Manage Risk Differently," Princeton University Press, 2007.

real world of business, we get to choose a strategic direction perhaps once every 5 to 10 years. (Of course, some organizations seem to change their strategic direction every other week, but that is a known and certain strategy for low employee morale and eventual organizational failure.) The real world of management is more akin to our coin flipping gamble where we only get to flip the coin once and the outcome is magnified. Few of us have a situation where we can grant 5,000 credit cards and rely on the statistical averages to even out among the various cardholders who turn out to be good or bad credits.

Assuming that statistics, and implicitly frequentist statistics, apply when the event size, or transaction size, is small is a serious abuse of statistics. The results can be disastrous, as disastrous as miscalculating the odds in our modified coin flipping gamble.

The second issue with frequentist statistics is that we rarely, if ever, know the true statistics of the underlying process. For instance, in our coin flipping gamble, we can reasonably assume that the coin is fair, and thus assume a 50 percent chance of turning up heads and a 50 percent chance of turning up tails. However, knowing the underlying distribution of events in managerial decisions is not a luxury that we usually have, and we, thus, need to make bold assumptions. For instance, it is frequently assumed that financial returns follow a normal distribution. The unfortunate reality is that realized returns are actually leptokurtic, which is a fancy pants way of saying that the odds of extremely positive returns or extremely negative returns are higher than the statistics of the normal distribution tell us they should be. Well, you say, that is an easy fix. Simply replace the normal distribution with the leptokurtic distribution and recalculate the statistics. Small problem though—we do not know how to calculate the statistics of a leptokurtic distribution. We, thus, revert to using the normal distribution that approximately works and then claim amazement when we get financial bubbles and our mathematics appear wonky with hindsight.

The third issue with frequentist statistics is that it blinds us to paradigm shifts. Consider, for instance, the 2008 financial crisis that in large part was based on a bust in the housing markets. Using the frequentist statistics available from the historical data available at that time, it was quite legitimate for the quantitative analysts at ultraconservative insurance company AIG to state that there was basically no chance that they could ever lose money trading securities based on housing defaults in the United States. A simple analysis of the historical data would have confirmed this. Of course, AIG almost went bankrupt and had to be bailed out by the government to prevent a financial collapse that threatened to take down a significant portion of the U.S. financial sector. What the highly trained analysts at AIG failed to appreciate was that a paradigm shift in the housing market was occurring and that, indeed, people in large numbers would start defaulting on their personal home mortgages with an unprecedented frequency. By definition, paradigm shifts are hard to predict, and thus easy to overlook or ignore when utilizing the convenience of a frequentist statistics assumption.

A final issue with frequentist statistics is the problem of distinguishing between a systemic event and a distinct event. This is an issue that is closely related to an event not being repeated frequently enough for the technique to apply. To illustrate, assume that you are a participant in a seminar that I am conducting. There are about 200 people in this seminar, and thus you feel comfortable in applying frequentist statistics. As I am talking, and as you are nodding off due to boredom, I suddenly drop dead at the front of the room. While you might be somewhat shocked, you likely will not be worried for your own well-being. For starters, you know that in a given room there is a nonzero probably that someone will drop dead, and it is better for it to be me who drops dead and not you. Also, you might surmise that my love of pizza has caught up with me, but with your superior diet of healthy salads that you have little to worry to about. You quickly conclude that my dropping dead was an idiosyncratic event. However, about 20 seconds later, someone else in the front row drops dead. And then another person and then another and soon almost everyone sitting at the front of the room is dead. Now your analysis of the situation changes from my love of pizza idiosyncratically killing me to one of a systemic effect at work, such as poisonous gas being dispersed throughout the room through the air conditioning vents. You correctly start to run out of the room for some fresh air and you abandon your frequentist statistical thinking.

When using frequentist statistics in the context of risk management, we rarely have the luxury of knowing if our use of frequentist statistics is valid or not until well after the fact. By the time the realization sets in, it is much too late to change our analysis and reverse our decisions based on that analysis.

Frequentist statistics is very tempting. It gives an illusion of precision and predictability. It allows for a wide variety of well-known statistical techniques to be used. It makes the analyst appear smart and knowledgeable. Frequentist statistics makes the effort exerted staying awake in statistics classes seem to pay off. It is true that it is very useful for those cases where large number of events with known statistical processes or distributions apply. The problem is that it can be very tricky to understand when it can be used with confidence and when it should be avoided. The differences between the instances are subtle, just as they were for our coin flipping gamble. Too often, we forget that the validity of frequentist statistical analysis is a false axiom of risk management.

Mark Twain once famously said that "there are lies, damn lies and statistics." If the field had been developed in his day, I am sure that he would have added "and then furthermore there is econometrics and big data." Frequentist statistics is great, but only when the conditions for its use are understood and respected. Otherwise, it is for humorists like Mark Twain to scorn.

Measurement Error

Everyone probably fondly remembers the science lab in high school. Every step of the way you had to take measurements and your lab instructor continually cautioned you that nothing could be measured with exactness. Thus, you also had to record your measurement error at each step and then correctly propagate those measurements through your calculations. The tedious calculation of those measurement errors is why everyone who has the talent becomes a theoretical physicist, rather than an experimental physicist.[2]

---

[2] If you need anecdotal proof of this statement, refer to the television sitcom *The Big Bang Theory* where the talented physicist Sheldon Cooper is a theorist, while his presumably less-talented roommate Leonard Hofstadter is an experimental physicist.

When was the last time you saw a risk measurement made with measurement error? More importantly, when was the last time you saw those errors in measurement carried through calculation? I suspect, very few times. In risk, we have the false axiom that data, particularly financial data, is made with zero measurement error. For instance, we see the closing stock price is $50.38. We take this as a fact that it is $50.38 and not $50.45 or even $50.37. The issue though is that the stock price is currently fluctuating through time. The price of $50.38 just happened to be the price at the instance that we looked at it. All financial prices are constantly moving, and thus it is much more accurate to state it as a range, with measurement error, than as a fixed and precise number. In fact, there is probably more uncertainty in the exact value of the stock than there was in any of the measurements that you made in your high school science lab. The error of the fact is of course magnified when we realize that each of our risk measurements—for example, calculating the value of a large portfolio of stocks—is summed up.

Now, the astute stock portfolio analyst will argue that this is a false issue, as they almost always state the value along with a standard deviation. However, it is well known that standard deviations are notoriously unstable. When was the last time you saw a standard deviation number given with an error measurement?

If you think back to the mathematics of calculating errors from your high school science lab, you quickly realize that the net error range for most risk calculations is often going to be extremely large. Think, for instance, of all of the assumptions that go into something seemingly objective as pricing a share of stock. Academic finance tells us the factors are the future dividends, the growth rate of those dividends, and the appropriate discount rate that accounts for the riskiness of those dividends. Each of these three factors is inherently uncertain with a wide range of reasonable values. Yet, I have never seen an analysis that puts an error range on these values, or propagates those errors through the calculation. While it is true that a good analyst will publish a range for what they believe the stock price will be, the range they provide is almost always much smaller than the range that would have been calculated, given the known measurement error by properly propagating the known errors through the calculation.

Continue with the argument and consider the measurement of risks that need to be made on nonfinancial data. What is the measurement error in calculating the tolerance of a manufacturing machine? What is the measurement error in calculating the confidence of consumers? What is the measurement risk in calculating the potential destruction of an oncoming hurricane?

As risk managers, we need to especially consider the measurement risk of Rumsfeld's famous "unknown unknowns." The secretary of defense, Donald Rumsfeld was ridiculed soundly for stating that "...there are known knowns; there are things we know we know. We also know there are known unknowns; that is to say we know there are some things we do not know. But there are also unknown unknowns—the ones we don't know we don't know." It is a valid and profound statement, but one that risk managers too frequently try to avoid acknowledging.

What is the measurement of virtually anything in the future? (Remember that risk is concerned with the future.) No matter what definition of risk you decide upon, it includes some aspect of uncertainty. It is, thus, very ironic that measurement error is missing in action. The problem of course is that incorporating measurement error into risk measurements makes the field seem rather suspect. Far better, for the health of the risk management profession to assume, measurement error is unworthy of attention.

One particular example illustrates the point. The recovery rates of corporations after bankruptcy compiled by a well-known rating agency are reported in one textbook on derivatives.[3] The average rate of recovery turns out to be approximately 50 percent. However, the standard deviation of this study show that the standard deviation, or variation in the measurement is approximately 25 percent. If we take two standard deviations as our measurement error, that implies that the range for recovery after default is between zero and 100 percent! Incorporating the measurement error in this one instance, of course, makes the risk calculations interesting.

---

[3] John Hull, "Options, Futures and Other Derivatives," 8th Ed, Prentice Hall 2011.

Admittedly, I played a little fast and loose with the calculation in the previous paragraph. Also, it is a lot more difficult to incorporate measurement error into most risk calculations than it was in your high school lab experiments. However, the point holds that risk management as a field would be a lot less confident in their projections, but perhaps, a lot more accurate (but not as precise) and truthful if they regularly included measurement error in their calculations.

There is another major benefit to incorporating error measurement and that is the discipline it imposes in thinking long and hard about the errors in your measurement. There is value in seriously considering the limits of how well you can measure a value or know the true value of something. It is humbling, but it is also enlightening.

It is past the time to get rid of the false axiom of no measurement error. It is far more honest and more productive in the long term to embrace the old adage that it is better to be approximately right than to be precisely wrong.

Unfortunately, measurement error in risk management is implicitly assumed to be zero most of the time. This is a serious error and a significant false axiom in risk management. It is also highly ironic when one considers that risk is the discipline of uncertainty of the future.

Optimization

Have you ever had lunch with a colleague at work and something like this statement came up: "Can you imagine that any organization can be as screwed up as ours?" I bet, you have. At virtually any organization where two or more people believe they can talk freely, something like that statement will be uttered. I bet you said it, and I bet that the people who report to you said it about you behind your back. I know that I have uttered something like that many times, and I am highly confident that the people who have reported to me said it about me and the decisions that I have made.

Frederick Taylor, the father of scientific management, led us on a path of management that made us believe that virtually all organizational processes could be optimized. Furthermore, if you have been to business school or even just a management seminar, you have almost certainly

drunk that organizational best practices Kool-Aid yourself. Sadly, organizational optimization is a false axiom—a myth. The reality is at best
"satisfication"—a search not for the best solution, but an acceptable solution that does not totally suck—to borrow a phrase from the millennial
generation.

As will be discussed at length later in the next chapter, What Is Complexity, and in a number of other different contexts throughout this book,
organizations are not machines that conform nicely to something like the
laws of physics and engineering. Organizations cannot be, and are not,
optimized. They are not subject to a nice set of formulas for which some
mathematical derivatives can be taken to calculate the maximum or minimum. Nor can risk management be optimized. However, a quick survey
of financial advertisements playing during the weekend football games
will feature companies that explain how they will optimize your financial
future by optimizing risk for your personal preferences. It sounds so nice
and reassuring, but it is pure poppycock.

There is another problem with optimization. What exactly does it
mean to optimize risk? Does it mean that there is as little risk as possible?
That, however, would limit the upside risk, and no company wants to do
that (do they?) Does the optimization of risk mean that the upside risk is
maximized while the downside risk is minimized? Nice thought, but how
exactly do you pull that off?

Anyone who has taken a formalized course in decision making under
risk will recognize the various frameworks; minimax (minimize the maximum loss), maximax (maximize the maximum gain) as well as various
others. These, of course, are all mathematical guidelines or philosophies,
or taxonomies for characterizing a decision, and not hard-and-fast rules
where everyone will obviously agree on the method to use. Optimization
in risk is a philosophical choice, and despite being a subject of discussion under olive trees since the time of Socrates (and likely even before
that), no one has come up with a universally accepted way to optimize
philosophical decisions. This is especially so for risk where everyone, and
every organization, has a naturally different tolerance for risk, and thus a
different calculus for calculating the risk–return tradeoff.

Thus, we have a false axiom of risk management in that the unexamined assumption is that risk can be optimized. The reality is "satisfication"

not optimization. In other words, the reality is a word that is so imprecise that it is not even a real word is the modus operandi, and not the very technical and precise sounding phrase of optimization.

If It Cannot Be Measured, It Cannot Be Managed

Management guru Peter Drucker stated that "what gets measured gets managed." Probably, in no area of management is this truer than in risk management. It seems as if risk management is actually measurement management. I have to personally come clean and admit that I often fall into this trap myself. I regret confidently claiming in a Business school discussion that I believed that anything in business could be measured. I was young and foolish when I said it. The problem is that a corollary of this is all too often taken to be true as well. That false axiom is that "if it can't be measured it can't be managed."

Any seasoned and wise risk manager knows that there are quantitative risks as well as qualitative risks. There are risks that we have techniques to measure—even if those measurement techniques may at times be suspect—and there are risks that cannot be measured. Additionally, experienced risk managers realize that it is often the unmeasurable qualitative risks that are the more significant ones and the issues that need management the most and benefit from management the most.

It is nice to be able to measure progress, to be able to definitively answer the question of whether or not things are improving. However, there are lot of things in life that would be nice to have true that are not.

You cannot measure stupidity, absentmindedness, emotions, social movements, market complexity, politics, actions of competitors and suppliers, weather, acts of nature, Black Swans, acts of God, and a host of other common yet critical events. These are the types of events that have a much more significant impact on the uncertainty of an organization's operations than the measureable risks.

Just because something is not measureable with a realistic degree of precision does not mean that you should not at least try to do so. Furthermore, it most certainly does not mean that you should not manage them. However, this is a message that risk managers, and particularly regulators, seem to forget. The rise of the mathematician who understands measureable risk, and the demise of the experienced, yet less quantitatively educated risk manager, is a tragedy as well as a travesty of modern risk management.

Risk is closely tied with human emotion. The response to risk is at the individual level an emotional artifact. For an organization, it is a sociological artifact. Emotions, either individually or sociologically, cannot be reduced to a number, nor should they be. Even if it could be tied to a number, that number or value would be constantly changing as the context changes. The emotional nature of risk, however, requires more human input into its management, not less. It requires more empathy and less mathematics. Uber may have self-driving cars, but risk management should not also be in a similar fashion be quantifiable digitized, so that a bot or a computer makes the decisions—or at least not for the majority of complex risk issues—a point that will be expanded and clarified in the following chapter on complexity.

For the risk decisions that can be quantified, then it is fine, and probably preferable for a computer or a bot to make the decisions. However, risk managers need to develop the wisdom to distinguish between measureable risks and nonmeasurable risks and manage accordingly.

Related to the role of measureable risk, and also with the axiom of optimization, is the concept that answers should be calculable. With many risks, the answer is not calculable and the best that one can hope for is an answer of "maybe." Not exactly satisfying, but it is usually both a reasonable and realistic answer and needs to be recognized as such. A real problem is that regulators, and in particular financial regulators, as well as shareholders, senior managers, and board members will rarely accept "maybe" as an answer. In truth, the old risk saying of "it is better to be approximately right than precisely wrong" seems to have been forgotten.

As a graduate student in physics, I had one professor stress the importance of never starting a calculation until you intuitively know the answer. His reasons for making us do so also apply to risk management. Firstly, the professor had a learning goal in mind. By forcing us to intuitively develop an answer, we as students were forced to develop our intuitions, and thus our understanding of physics. Secondly, when the resulting calculation, or experiment, produced a different result, we knew that either we had a mistake in our intuitive knowledge, a mistake in the calculation, or in the best case scenario, we discovered a new phenomenon. By basing risk management on a precise black box calculation, all of these benefits are needlessly thrown away. The focus on measureable risk tempts us into such black box calculations and the learning unfortunately only takes place after painful and costly mistakes.

You are likely familiar with the serenity prayer by Reinhold Niebuhr. It goes, "God grant me the serenity to accept the things I cannot change, courage to change the things I can, and wisdom to know the difference." I think this should be changed to "God grant me the serenity to accept the things I cannot calculate, courage to calculate the things I can, and wisdom to know the difference." It is time to regain the plot and to start managing the risks that cannot be measured.

Noise

In 1985, the renowned finance academic Fischer Black wrote a paper that was presented at the annual meeting of the prestigious American Finance Association. The paper was simply titled "Noise."[4] In this very insightful thought piece, which has unfortunately been largely forgotten, Black highlights that too often what we take for meaningful data is, in fact, nothing more than meaningless noise.

The acting on noise, or perhaps more accurately confusing noise, for information leads us to calculating values and measuring risk based on false premises. The law of one price in finance states that similar sets of cash flows should have similar prices. Thus, there is a search for similar situations or arbitrages from which the value of a situation is calculated. The reality, however, is that false arbitrage is the norm. A false arbitrage is when you think something is similar when, in fact, it is not. For instance, every movement is the financial markets or in commodities prices is seen as significant and debated at end by the TV pundits. While the media needs to fill space and time, we must carefully filter the real data from the noise.

---

[4] Fischer Black, "Noise," Journal of Finance, Volume 41, Issue 3, July 1986, 529–543.

In the quest for measurement and quantification, there is a need (a lust?) for quantitative risk data. Frequently, however, noise is confused for data. The confusion leads to more problems than it solves. The problem obviously becomes even greater in this age when Big Data and the techniques for processing it are coming ever more popular.

Brain's Win

Leslie Orgel was a chemist and a biologist who studied the early origins of life. His studies in chemistry and biology led him to famously claim that "evolution is cleverer than you are." This also applies to risk management. Organizations are essentially an evolutionary system, as are industries, and as are economies. They are directly akin to biological systems in how they change and evolve. Thus, Orgel's law about evolution applies to organizations as well. The assumption, however, in risk management is that "brain's win." Namely that with enough brain power that any risk problem can be solved, or better yet optimized. The reality is that, just like all the king's horses, and all the king's men, risk management cannot put Humpty Dumpty back together again.

Intelligence, creativity, and intellectual effort certainly help in risk management. However, that does not necessarily mean that they win. In any type of complex evolutionary system, the power and the mysteries of evolution will dominate. Referring again to the hit television sitcom *The Big Bang Theory*, it is the dim-witted waitress Penny who generally succeeds, while the superintelligent physicists bumble through life as their hyper-rationalism and knowledge continually trips them up in their everyday real life activities. While brains are useful, and necessary, hyper-rationalism rarely succeeds when human emotion is involved. In risk, human emotion is always playing a role.

Concluding Thoughts

While we may not have completely lost the plot in risk management, I do fear that we are well along that path. Risk management needs to be based in real life. It needs to be based in human emotions, and it needs to be based in complexity, which is the subject of the next chapter. Relying on axioms has some significant limits. Axioms make risk management seem like more of a science, and mathematically based axioms make risk management easier to deal with. However, relying on false axioms does not make risk management more realistic or effective. Realizing when these axioms apply and when they do not is key to successful risk management. Without these realizations being made, risk management truly is in danger of losing the plot.

CHAPTER 3

What Is Complexity?

There is perhaps no concept more critical for the risk world to gain an appreciation and an understanding of than the difference between things that are simple, complicated, and complex. These are all types of "systems" that systems thinkers use to categorize problems or issues or situations into. They are relatively easy to distinguish, and any experienced risk manager should be able to easily relate to the differences. The categorization though is not an academic exercise, as the different type of systems needs to be thought about and managed very differently.

There are two problems though. The first is that there is little knowledge about systems thinking, and thus very few risk managers are even aware that these different ways of thinking about systems exist. This lack of awareness is true on both an explicit knowledge level, and perhaps, more troubling there is not even an understanding on an intuitive basis. The second problem is that the risk management field has complicated thinking tools, and we all know the old adage about having a hammer and then acting as if everything is a nail.

As will be argued in this chapter, complicated thinking is rarely appropriate in that vast majority of important risk decisions. Making the problem worse, complicated thinking has the potential to make the solutions and actions devised cause much more harm than doing nothing. Indeed, it can be argued that the 2008 crisis was caused almost entirely by complicated thinking. Furthermore, many pundits, including me, believe that the continued insistence on complicated thinking is only going to make the next subsequent crisis not only more likely, but also more severe.

Many of the false axioms discussed in the previous chapter are the result of complicated thinking. Complicated thinking is convenient, and it is also, in a way, sexy. As the name implies, it is frequently complicated in the normal sense of the word, and thus, it lends credence to risk as a credible profession and specialty. Complicated thinking has its uses, and indeed, is a valuable and necessary skill for the risk manager to have—but only for those situations where it is applicable. Medicine is generally a complicated profession, but one does not want a heart surgeon to be running a sophisticated currency hedging program that utilizes exotic derivatives, just as you do not want a risk engineer to perform your heart bypass surgery.

It is time for the risk management profession to learn about systems, and more specifically, about complexity. The purpose of this chapter is to introduce systems thinking, illustrate the importance of it, and put complexity thinking at the forefront of risk management as it should be. Complexity explains a lot of different business and economic phenomena such as why certain videos go viral, the rise in importance of social media, the emergence of fads and even boom and bust cycles in the stock market. Complexity can also help us understand a lot of issues concerning risk. More importantly though than explaining things, complexity provides a new way of thinking about how to manage risk. A paradigm shift is needed in order to move away from the ever increasing dominance of complicated thinking to the more nuanced and appropriate complexity thinking.

Ecology for Bankers

In the 1970s, noted biologist Robert May started to warn the profession of biology that there was something else behind natural processes than the objective and reductionist rules that biologists were applying to advance their field as a discipline of scientific study. That phenomenon that Robert May was warning of was the nascent field of complexity theory, which itself was a subset of the larger systems way of thinking. The field of risk management should revisit Dr. May's arguments, which proved to be invaluable for progress in understanding biological processes.

Dr. May, along with Simon Levin and George Sugihara, wrote a short, but very interesting article titled "Ecology for Bankers."[1] The article

---

[1] Robert May, Simon Levin, and George Sugihara, "Ecology for Bankers," Nature, Vol. 451, 21, 893–895, February 2008.

highlights the similarities between ecological systems and systemic risk in the financial system. Ecological systems are connected as are financial systems. Connected systems where each of the parts can adapt to the actions of other parts of the system are a type of system that is known as complex. A complex system reacts in very particular and unpredictable ways. It is directly opposed to what is known as a complicated system, which has parts that act independent of how other parts of the system act.

The predominant paradigm, particularly in risk management, is that the world is rational and reductionist. In more basic terms, this implies that things work according to a set of laws, and furthermore, that each of the pieces of a system or an issue can be looked at individually and then put together to form a logically consistent whole. That this dominant paradigm is particularly obvious if one examines the current academic research in risk management.

Studying phenomena with a completely rational and reductionist mindset is a clockwork view of the world. It assumes that outcomes are reproducible, in that if one conducts the exact same activities, then one will get the exact same results. However, risk management has several important differences from the workings of a clock. A clock is composed of springs, gears, and levers, which work in precise and predictable ways that are based on the well-known and well-tested laws of physics. In order to create an accurate watch that will last for decades, all one needs to do is put the appropriate pieces together in the correct sequence and in the correct proportions according to a set of blueprints. Risk management, meanwhile, has no mechanisms that work in precise and well-known ways. You cannot put together a successful risk management program that will last for decades by simply following a set of blueprints.

The clockwork view of the world is the basis of the scientific method that we all learned in primary school. Business, of course, also has its equivalent in scientific management that was first popularized by Frederick Taylor. The scientific management paradigm is, of course, the reason behind business schools and Masters of Business Administration programs. Scientific management is also the basis for most of the discipline of risk management. It is a rational, rules-based view of risk management, which assumes that risk management is a scientific discipline that operates much the same as physics does. It is a very desirable paradigm for several reasons, but it is also patently false, misleading, incomplete, and an arguably dangerous paradigm to be operating under.

Introduction to Systems Thinking

Systems theory is an evolving paradigm that is proving to be of great use in providing a different lens in which to examine issues in the management of natural systems, social systems, information management, and in general, business management itself. The qualities of systems thinking make it particularly appropriate for risk management.

The primary characteristic of systems thinking is that the assumption of reductionist thinking being the dominant paradigm should not be the automatic assumption. The ability to break a system down into its constituent parts and develop an understanding of the system simply by examining each of its parts, and then reconstituting the parts to create the whole is seen as an incorrect and misleading way of analyzing situations, unless it can be clearly demonstrated otherwise. Systems thinking involves examining the underlying processes that affect how a situation unfolds in a holistic fashion.

The three main types of systems are simple, complicated, and complex. Something that is simple is something that can be accomplished by following a few simple rules or, if you will, a recipe of sorts. It usually does not take any special training to be able to deal with. Furthermore, it is not particularly sensitive to how closely the rules or the recipe is followed. For instance, conducting the safety announcement on a plane is an example of a simple risk task.

A complicated system or task is a process that follows very well-defined and rigid rules or laws. It is also a process in which the results are reproducible in that if the exact same conditions exist, and the exact same steps are followed, then the exact same results will be produced. A complicated process is often compared to Newtonian physics in which there are well-known rules from which very precise outcomes, such as the movement of the planets, can be calculated with an amazing degree of precision. An example of a complicated risk process is dynamic hedging of a financial option by using the Black–Scholes option pricing formula.

If one uses the same inputs, one will get the same hedge ratios of the offsetting position, and the same profit or loss from hedging the option if the underlying price process is the same.[2]

Simple and complicated systems share the feature that they are rules-based. The difference is in the robustness with which the rules apply. The presence of rules, even loosely applied rules, means that managing simple or complicated processes are relatively straightforward, once one knows the underlying rules or recipes. Underlying a simple or complicated process is a known—or at least conceptually knowable—process or set of processes. Additionally, simple and complicated systems can be analyzed in a reductionist fashion; their component parts can be broken down, studied and managed separately, and reassembled to construct the whole. This makes management of them a relatively straightforward task.

A key aspect of simple and complicated tasks is that they can be digitized, and thus managed by a computer or a robot. For instance, self-driving vehicles, particularly on the highway, is an example. Self-driving vehicles are projected to be much safer and efficient than human-operated cars. As computing power increases, and as the mechanics of robots improve, many more tasks can be expected to be automated, which, in turn, increase the safety and efficacy of many tasks.

The roots of scientific management are rooted in the reducibility of complicated systems. Take, for instance, the assembly line that Frederick Taylor is credited with inventing and Henry Ford is credited with perfecting and proving the value of Taylor's famous time–motion studies in which he broke down each single step of a task into its component parts, and then individually maximized each component was key to development of the modern factory. For instance, Taylor went so far as to measure the height and placement of the assembly line in relation to the worker to make the actions of the human assembly line worker as

---

[2] Ironically the principles behind the Black–Scholes Merton model for dynamic hedging were used by the hedge fund Long-Term Capital Management, which was perhaps better known as LTCM. LTCM, which had as two of its partners the Nobel Prize-winning economists Myron Scholes and Robert Merton suffered losses that led to its bankruptcy due to its strict adherence to the use of the formula. It is a striking example that risk management is not complicated.

compact, time-efficient, and energy-efficient as possible. As an example, Taylor's studies demonstrated that the ideal shovel size would allow a man shoveling to move exactly 22 pounds with each scoop of the shovel. This would lead to the most material being shoveled within a workday. Thus, the shovel head size was optimized to allow for 22 pounds of material. Different materials with different densities would necessitate different size shovel heads.

A complex system is fundamentally different. A complex system is one in which a phenomenon known as leaderless emergence is a central feature. In a complex system, there are no underlying rules or processes. Although patterns may be observable, they are not predictable, nor are they repeatable. Doing the exact same thing will not produce the exact same results. In fact, the results are just as likely to be totally opposite from previously.

Complexity arises whenever a situation involves a number of different agents that have the capability to adapt and change. In essence, complexity can arise whenever there is a group of people. A classic example of a complex system is the stock market. There is no leader of the stock market. Although there are many stock market pundits, none of them control whether the market rises or falls over any given period. Whether prices go up or down depends on the collective actions of the thousands of individual traders in a given stock. In turn, each of these traders will adjust their actions and their perception of the correct stock price based in part on the actions of other traders. It is a set of continually adapting actions that leads to the prices of stocks rising and falling. Stock market patterns are observable with hindsight, but they are not predictable. No one knows at any given point of time whether the market will rise or fall or remain relatively calm and unchanged. It is a dynamic system, and it is an incalculable system. It is complex.

Where Does Risk Management Lie on the Spectrum?

In my experience, almost every risk management problem of consequence falls into the realm of complexity. The problem is that almost every risk management response is a complicated one. Thus, there is a fundamental mismatch.

As will be discussed in the next chapter, major risk management issues almost always involve people, either as individuals, groups, societies, or even economies. Rarely is a major risk management issue solely a mechanical one. Even if it is a mechanical one, the response is generally straightforward—for instance, if a part in a machine breaks, then you fix or replace the part. If a formula in an algorithm is not doing its intended function, then you reprogram the algorithm. However, you cannot so easily replace or fix a person and you cannot reprogram a person or a society or an economy.

As discussed earlier, complexity arises when is are a set of agents that interact and that can adapt. The business world, and the global economy, is exactly that; a group of agents (customers, competitors, employees, vendors, investors, regulators, and so on) who consistently interact (sales transactions, advertising, social media, news feeds) and adapt (change strategies, develop new products, discover new wants and needs, change interests, change their opinions). Business and the economy are complex, and thus by extension, risk management is complex.

The possibility of bad or good things happening is not something that is designed by a complicated formula. If it was, I would hope that business leaders, economists, politicians, and central banks would have figured it out and programmed the economy to work in a more efficient, predictable, and benevolent manner. They have not simply because the world of business and the economy are complex entities. That is the central reason why economics is called the dismal science; it is not a complicated science at all!

Once you begin to think about it, and once one understands how complexity arises and the implications of complexity, it appears to be obvious that risk management is for the most part a profession of managing complexity. Despite the obvious, however, risk management is dominated by complicated thinking. There appears to be three main reasons for this: (1) an ignorance of complexity due to the relative newness as well as the novelty of the discipline, (2) a momentum built up from the engineering roots of risk management, and (3) an unwillingness to change a way of thinking based on arrogance and a false sense of confidence. Adding to this is the current age of the well-educated and highly degreed and certified expert—what I call the "white coat effect," which is if someone is thought to be a scientist of any sort, then it is almost automatically assumed they will be an expert with readily available answers and solutions.

Outside of the social and the biological sciences, there is an overall ignorance about complexity. It is rarely discussed in business schools, nor is it discussed in most advanced courses in risk management, which are based on mathematical principles, (that is, the complicated-based principles), of data analysis. In my discussions with experienced risk managers about complexity, they immediately get the concept and understand its appeal and suitability as a working model for risk management. However, I have had many well educated, but less experienced risk managers who argue ferociously against a complex view and instead put forth their complicated ideas. Complexity is seen by them as a threat to the acceptance of risk management as a discipline worthy of respect.

This brings up the second issue of a momentum built up from the engineering roots of risk management. There are a lot of risk managers, and risk management organizations, who have a strong vested interest in keeping the status quo based on complicated thinking. It is extremely difficult to test someone's ability to deal with complexity, but rather trivial to test someone's knowledge of complicated things. Thus, schooling is based on complicated issues, certifications are based on complicated issues, and regulators institute processes based on complicated thinking.

With complexity management, you manage a situation, and you need to realize that it is neither possible nor reasonable to believe that you can solve a complex issue. This thinking of course goes directly against an engineer's view of the world, where every issue is a problem waiting to be solved if only enough brainpower is applied to the problem. The "manage, not solve" paradigm of complexity management also requires a huge amount of humility. For any highly trained individual, such humility may be difficult, if not impossible, for them to live with.

This brings us to the third issue behind the dominance of complicated thinking even though complexity is the more appropriate mindset. Everyone has an innate need to believe that they are needed in their organization, and there is no reason to think that risk managers would be any different. Having a complexity mindset, where the answer to any complex question is generally of the form "maybe," is not an answer that many managers have the self-esteem to feel comfortable giving. Risk managers want to believe that they are "Masters of the Universe" to borrow a phrase from author Tom Wolfe.[3] They want to believe that through their skill, understanding, and intelligence that they are capable of controlling outcomes. The reality is quite different.

How to Manage Different Types of Problems?

The different type of problems—simple, complicated, or complex—obviously require different types of management techniques and tactics. In fact, attempting to manage one type of system, say a complex issue, with simple or complicated type tactics is likely to cause far more harm than good. At a minimum, using the wrong management tactic is likely to be inefficient and ineffective.

The simplest type of system to manage is not surprisingly a simple system. Because a simple process can be best managed with a recipe or a guide book, the obvious thing to do is to use the recipe or guidebook. To take a very common simple task, consider packing for a trip. I travel frequently on business, and to expedite the packing process, I have actually made up a business trip packing checklist. While it may seem like a very unnecessary thing to do, it is amazing how having a checklist both expedites the task of packing, as well as reduces the risk of you winding up in a city far from home and realizing you forgot cufflinks that you need for an early morning breakfast meeting. In fact, I once had a meeting with someone who had the hotel staple their cuffs together, and someone else who used paperclips as a substitute for their forgotten cufflinks.

A checklist provides a very simple, but also a very effective risk management tool. With my simple business packing checklist, it not only reduces the risk of forgetting something, but it also reduces the amount

---

[3] Tom Wolfe, "Bonfire of the Vanities," 1987, Farrer Strauss Giroux. The novel concerns a successful bond trader who believes that through his intelligence, he alone is responsible for his impressive success; that he is a "Master of the Universe." Through a series of subsequent events, he learns to his chagrin that he conversely has little control over his life. It is a fictional example of how complexity lays waste to the best laid of plans and intentions.

of thought and stress that needlessly goes into packing. There are none of those pauses where you ask "Have I packed everything?" as you stand there looking around your clothes closet or your office while scratching your head.

In a very interesting book, physician and author Atul Gawande outlines the advantages of using checklists and their usefulness in medical contexts.[4] He gives clear evidence of how the use of simple checklists in the preparation for a surgical procedure dramatically increases the probability of success. It may seem so trivial that it is almost unnecessary, but the evidence is clear. Of course, medical procedures have a much greater level of risk and a much greater level of significance than my packing for a business trip.

Checklists are ideal for the risk management of simple systems and even complicated systems. However, caution should be used, as checklists tend to be very limiting when dealing with complex systems. A checklist is only appropriate if the task at hand is truly simple and routine. If there is the possibility for complexity to emerge, or for a nonroutine component to become a factor, then simply using a checklist may be leading to blindness to the possibility of nonroutine factors. For instance, checklists can lead one to a false sense of confidence, or lull one into a routine. If something nonroutine happens, it can be easily missed.

When dealing with a complicated system, the correct response is, of course, to engage an expert in the particular process or to follow the associated rules and laws. As they are completely defined and governed by laws or rules that produce completely reproducible results, complicated systems might actually be the easiest to risk manage—assuming, of course, that one knows the underlying laws that govern the system. Computers and robots are the ideal managers for complicated tasks, as they always produce the same results or outputs given the same inputs, and they respond more quickly and are less prone to error than humans. To reiterate, however, few systems or problems faced by organizations are completely complicated.

---

[4] Atul Gawande, "The Checklist Manifesto: How to Get Things Right," 2009, Henry Holt and Company.

There are three main steps for managing complexity. They are: (1) recognize whether the issue at hand is simple, complicated, or complex, (2) think "manage, not solve," and (3) engage a "try, learn, adapt" approach.[5]

The first step in managing complexity is to recognize that the system is complex. This, by itself, will help to prevent automatically resorting back into complicated thinking mode. The mere recognition of this fact will help, in that a more flexible approach will be utilized than usual.

The second step in dealing with complexity is to think in terms of "manage, not solve." The phenomenon of emergence, which is a fundamental outcome of complexity, means that one has to be humble in one's expectations. A search for a solution will prove futile. Complex systems are too unpredictable and involve too many moving parts for one to realistically believe in a set solution. Unlike with complicated systems in which the issue can be decomposed into its constituent parts and each of the parts managed in isolation, that is not possible with complexity.

"Manage, not solve" is very difficult conceptually for the risk management profession to follow. It goes against the engineering background of the field, it goes against the quantitative analysis bias of the profession, and it requires an almost inhuman amount of humility to admit that there is no human-based analytical solution available. It is also taxing and difficult to do. When you solve a problem, you are done with it. When you have to manage an issue, however, you can never let your guard down; you can never stop thinking and adapting to the evolving situation. While complicated issues can require a large amount of brainpower and energy to solve, they are in the long run far easier to deal with than complex issues, which are indefinitely ongoing.

"Manage, not solve" is also difficult to do, given the pressure that risk managers are under by regulators, boards, senior managers, and even the general public to provide answers and concrete solutions. The expectations of the various stakeholders of an organization are rarely realistic, given the presence of complexity. It creates a quandary for the risk manager that can only be partially addressed with education and communication of the

---

[5] This three-step process for managing complexity is adapted from the book "It's Not Complicated: The Art and Science of Complexity in Business," by R. Nason. Forthcoming 2017, University of Toronto Press.

issues surrounding complexity management. It requires a luxury of time, patience, and understanding that risk managers are rarely ever given.

The third component of managing complexity follows from the "manage, not solve" mindset. The third aspect is to adapt a "try, learn, adapt" mode of operation. The fundamental characteristic of complexity is emergence, which means that the risk manager must also practice emergence. The way, of course, to do this is to try things, see how the system reacts, learn from the reaction, and then adapt accordingly. Emergence comes from continual adaptation of the parts of the system. Emergence, thus, necessitates that the risk managers also continually adapt. It requires the risk manager to keep not only a keen eye, but also an open eye (and on open mind) on how the complex risk issue is evolving. "Try, learn, adapt" is active, not passive risk management. It involves the risk function becoming a living, breathing, and an in-the-moment function, just like the risk itself is. It is the antithesis of the analyze, plan, implement, and let the plan take care of things approach that so often is de facto utilized.

Recall that complexity arises from a group of agents who interact and can adapt and change. By its very nature then, complexity management requires a dynamic, rather than a static approach. With the "try, learn, adapt" approach, it is not so much the knowing of tactics that it is important, but the creativity of the tactics. You need to try things. You need the flexibility to actually make mistakes. You need the humility to realize that you cannot always know the answer a-priori, and furthermore, you may not know the answer ex-ante. You need an open mind to learn intuitively, rather than textbook learning, where the new idea becomes codified into a new process. After all, you must remember that complex systems are not reproducible like complicated systems. Thus, what worked one time is not likely to work the next time.

Risk management in the face of complexity necessitates that the tools of risk management need to be flexible and adaptable as well. Quarterly risk metrics or annual reports will not produce the timeliness needed. However, an organization that is itself steeped in the awareness of, and that has the willingness and the flexibility to adapt complexity risk management, will itself be an emerging "machine" that will likely naturally adapt as necessary to a variety of emerging risk management issues in the timely manner necessary.

VUCA

It seems natural to conclude this chapter with a brief introduction to the phrase VUCA. VUCA, or volatility, uncertainty, complexity, and ambiguity, is a natural way to extend the concept of complexity. As a concept, VUCA has been embraced in a variety of contexts, but perhaps nowhere more concretely and successfully as in the U.S. military.[6]

If anything, risk is all about volatility, uncertainty, complexity, and ambiguity. However, the problem is that conventionally risk management systems, and the training of risk managers, has been almost exclusively about the first two components, with an almost deliberate ignorance about the last two components of complexity and ambiguity.

VUCA, however, is a paradigm that fits quite well into the techniques of systems thinking. The volatility and ambiguity parts are essentially the basis of simple and complicated thinking. Complexity, however, has been relatively ignored for the reasons previously discussed. It is time for the risk management to adopt VUCA, and in particular, complexity concepts and accept the humility that doing so involves.

Concluding Thoughts

There is a final, and perhaps the most important, aspect to managing complexity and that is to embrace complexity rather than fear it. Complexity is a fact in business and it is a fact in risk management. However, complexity does not just pick on one company or one organization—although at times it may seem like that. Instead, complexity is like playing tennis in a strong wind. No one likes to play tennis in a strong wind as it upsets a player's timing, their strokes, and a player's game plan or strategy. However, the wind affects both players. It is the player who adjusts the best who wins, while the player who moans the most about the conditions and how they cannot use their normal game

---

[6] A more extensive introduction to VUCA can be found at R. Nason and E. Mare, "The Need for VUCA Management in Finance Education," Technical Report UPWT 2015/24, Department of Mathematics and Applied Mathematics, University of Pretoria, 2015.

and strokes almost always loses. The risk manager who acknowledges complexity, and works with complexity, and even figures out a way to exploit complexity, is the risk manager who will prove their worth to their organization.

Philosopher Eric Hoffer once stated that, "in times of change, learners inherit the earth, while the learned find themselves beautifully equipped to deal with a world that no longer exists." In risk management, our problems are becoming more complex, while the simple and complicated issues are becoming less important (or being managed by a "learned" computer or robot). It is time for risk managers to become learners of complexity.

CHAPTER 4

What Causes Risk?

It is generally assumed that if you can get to the root causes of something, then you are well on your way to mastering it. So, that begs the question of "what cases risk?" It is ironic however that this is rarely if ever asked proactively. Instead it is almost always asked after the fact, as in "what caused this mess?," or "what caused this screw-up?" Notice as well that it is almost never asked after something good happens. Rarely does someone seriously ask, "How were we able to sign this customer to such a large deal, or why did our advertising go viral?" However, we must also remember that there is also upside risk, and in our analyzing the root causes of risk, we must not shortchange it.

What causes risk is a more profound question than it may appear to be on the surface. A quick response would likely, and reasonably be, that there are many causes of risk. I, however, believe that for the situations that we care about most, that is, for the risk situations that have the biggest positive, or the biggest negative impact, that there are two main causes that are almost always present—people and complexity. Furthermore, it might be argued from a close reading from between the lines of the previous chapter that it is people that cause complexity, and thus all risk, good or bad, is caused by people. Yes, there are other risks that exist—material breakage, acts of God, and randomness along with many others—but most of the time when something unexpectedly bad or good happens, there is a person at the root cause of it.

This chapter explores the idea that people and complexity are the root causes of risks. In order to buy into this somewhat novel thesis requires one to rethink many risk management activities. Risk management policies that are not grounded to managing root causes are likely to be both misguided and ineffective. Currently, most risk policies are seemingly designed for random events or acts of God or for complicated processes that we believe we have the mathematical tools to deal with.

Understanding that people and complexity are the roots of almost all of the risks that matter requires a significant change in thinking and in risk management practices.

Problems with the Question "What Causes Risk?"

What causes risk in your organization? It is a great question and an obvious starting place for developing a great risk management function. Knowing the root causes allows one to work from first principles. It allows risk to be mitigated or exploited at its core. Although asking what causes risk is a simple question, there are a few problems inherent in the question. To start with, there is an inherent bias in the question for the past tense. Risk is a future-looking event. However if I ask that question at your organization I will likely get a laundry list of all of the bad things that have happened in the past.

There is a better set of questions to ask. The first, and less valuable, is what has repeatedly caused risk in your company, and the second, and more interesting, question is what will cause risk in your company? The question about the past is interesting, but it is incomplete and not always that useful. The question about the future is much harder to answer, but much more interesting and potentially much more valuable to have the answer to. Risk is something that happens in the future. There is no uncertainty associated with the past. Thus asking what happened is a bit of a fruitless exercise with the exception of those cases where the adage of those who ignore history tend to repeat it applies.

Thus the first problem of asking about what causes risk is to get past the historical analysis of risk to the future analysis of risk. What caused risk is only valuable for how it will help manage risk going forward. However this leads to the oft-cited and frequently correct criticism of risk management that it is always working to fix the last crisis without enough focus on the next problem or opportunity.

The second inherent bias in the question of what causes risk is the normal risk biases of assuming that risk is the possibility of something bad happening. Now in an organizational context, if something unexpectedly good happens then there is generally a person, or more likely persons, who are more than willing to take responsibility for the unexpectedly good that happened. Saying that people are the root cause of good risks is a trivial thing to do as it is trivial to find management and staff at any organization that are more than willing to take credit for almost anything positive that happens. Thus something unexpectedly good is not seen as a surprise, or something unexpected, or as a risk, but as the product of conscious and "brilliant" foresight of a member or members of an organization. While it might very well be true that it was indeed the brilliant actions of a person or a group that caused the unexpected upside, simply saying it was brilliance and not a risk worth analyzing is not helpful. It certainly does not help the organization increase the odds of a good risk happening to them again.

This leads us to a third problem with the question of what causes risk in that it starts to assign responsibility for the risk. Now as stated in the previous paragraph, it does not take much effort in getting people to accept responsibility for a good risk event. However assigning responsibility for a bad risk event is quite different. Assigning responsibility for a bad risk is akin to assigning blame. No one wants to take blame. Very little positive comes from assigning blame in a risk context. It is important to realize that no one gets up in the morning thinking they are going to be in an accident that day. Likewise, rational people do not get up thinking they are going to win the lottery. Assigning blame only serves to make someone feel worse, who very likely already feels awful about the mishap they caused. (By the way, if people are actively trying to create bad risk events, or do not care about good risk, then the company does not have a risk management problem, but a very serious problem of a very different organizational cultural basis.)

The role of responsibility, blame, and risk management will be discussed at ore length in Chapter 8 when risk culture is discussed, but at this point it will suffice to say that the purpose of asking the question "What causes risk?" is most definitely not to assign blame.

Five-by-Five "Whys?"

If you are at a progressive risk management company, you likely have a regular report on risk events in the past. Depending on how seriously your company takes risk, your company may have even followed up and looked at the causes of those risks. However did they look at those causes superficially or deeply? Furthermore did they just look at negative surprises or did they also examine the upside surprises?

Some companies are excellent at following up on the root causes of risk, while others just categorize and assume that someone did not follow the stipulated process properly or took a shortcut. That may very well be true, but then one should ask the follow-up question of why they did not follow the appropriate process. That is a much more valuable piece of information to have. The first and easy answer is generally not very helpful nor valuable. It is also lazy thinking, and risk management cannot afford lazy thinking.

Therefore assuming that people are consistently skirting the rules is not really constructive analysis. Likewise assuming that employees do not care about risk is also incredibly lazy and cynical thinking.

Using a technique I call a "Five by Five Why" analysis is one way to get to root causes that prevents a superficial conclusion. In a five-by-five "Why" analysis, at least five reasons for the risk event are listed. Note that it may take some creativity to get to five possible reasons, but there is value in forcing one to come up with five different possibilities. Then, for each of those five "Whys," there are five questions or five "Whys?" asked that follow from each other.

The power of this method was clearly shown to me when I was talking to a student who came to me to discuss their career. I asked the student what they wanted to do with their career, and they mentioned that they wanted to have a career like a well-known money manager. With that response I started the whys. I asked the student for five reasons why they wanted a career like this certain money manager. They gave me five "whys," the first of which was that the money manager was rich. With this, I started down the path of asking five whys on this first response. I asked the student why they wanted to be rich. They responded "so I can buy the same fancy car that they bought." I then asked why they needed to have such a fancy car, and they responded "so I can take lots of friends for drives in it." I then asked the question "Why do you want to take lots of friends for drives in this dream car?," and they responded "because I am lonely."

This story is more than a little sad, but it illustrates that the first response you get to the question "why" is generally very far from the real truth, and very far from one of the root causes. I never would have guessed that the reason why this student wanted to go work for a major money manager was solely to get rich in the false belief that they might be able to buy friends. Furthermore I suspect that the student themselves did not understand their true motivation. When the student said that they wanted to work in money management in order to get rich, it would have been trivial to jump to the conclusion that that this student is just a stereotypical greedy MBA student and be done with the conversation. Asking the five-by-five whys however led to a much more profound conversation. It turned out that the student did not like finance or money management in the least. Going to work in money management would have likely have been a terrible career choice, but without getting to the root causes of the actions, decisions, or motivations makes it hard for anyone to understand what is really the motivating factor.[1]

In risk analysis I have seen too many cases where the analysis ends with the first obvious, yet incomplete answer. The first answer, which is generally the superficially obvious cause, generally has many more layers to it. The first answer is almost never the root cause. Consider for instance the loans to NINJAs (no-income, no jobs, no assets) that bankers were widely criticized for making in the build-up to the 2008 financial crisis. It is widely assumed that the bankers were just greedy in making those loans. However it has been my experience that bankers are very loathe to lend to those they think may not pay them back. So did anyone ask why the bankers all of a sudden became so greedy as to make those questionable loans? (Could it have been the regulatory lending requirements of the Fair Housing Act, or could it have been that they knew they could sell those loans forward? If, so then you need to ask what factors were in place to force the Fair Housing Act, or that allowed them to sell those loans to others and so on.)

Utilizing the five-by-five whys greatly helps to move beyond the superficial into a much more productive analysis and understanding. Even if the first and obvious answer does turn out to be the root cause,

---

[1] Needless perhaps to say, but a continuation of the five-by-five whys ceased at this point and the discussion went in a different direction. The student is now successfully working in marketing and has a wealth of friends.

going through the five-by-five analysis is valuable in that it exposes other potential causes which in turn will lead to a greater likelihood of them being mitigated or exploited in the future.

It is also important to note why you need to come up with five reasons. There is a very real temptation to go with the first cause that comes to mind. Often the first cause is a true cause, but almost as often it is not a cause but more of a symptom. Even if the first blush cause is a cause, there are likely to be at least one or two more enabling causes. Risks, both good and bad, tend to happen because of a confluence of events, not a singular cause. Additionally, forcing oneself to come up with (at least) five causes, forces one to be creative and to think beyond the usual suspects. There is a tremendous side value in doing this as it helps to build the risk intelligence and the risk awareness of the firm. Recall my first law of risk management; the mere fact that you acknowledge that a risk exists automatically increases both the probability and the magnitude of it occurring if it is a good risk, while simultaneously decreasing the probability and severity of it occurring if it is a bad risk. There is real value in forcing oneself to be creative when it comes to risk.

In an engineering context, we think of materials failure. Materials failure is relatively easy to diagnose. You see the broken pipe, or you see the stress fracture. People failure is much more difficult to assess. You see that someone did not follow protocol, or you see that they made a mistake. What you don't understand though is what caused them to not follow protocol, or what caused them to make the mistake. Again, only the psychotic go to work each day intending or believing that they will be the cause of a screwup. They might be afraid of doing something incorrectly or of causing a mishap, but that is not their intention. Given that, it is thus very superficial to say that the problem is that someone did not follow protocol or someone made a mistake. That is simply lazy analysis.

People People People

In real estate the old adage goes that the three things that matter are 1. location, 2. location, and 3. location. In risk management the three things that matter are 1. people, 2. people, and 3. people. Organizations are run by people, for people, and are all about people. I have yet to see an advertising campaign aimed at selling to computers. Robots have needs, such as batteries or new gears, but they do not have wants or desires, and thus consumerism is out for them. Inanimate objects do not browse catalogs wondering if they are going to be in style for next season. Computers do not have hopes and fears and thus emotional appeals to them are nonsensical.

People on the other hand do have wants, desires, needs, hopes, fears, and emotions. Risk management would be a lot easier without people. In fact, risk management could be, and should be, run by computers if it were not for people. One of the curious things about the debate of driverless cars is how people are afraid of them, yet many of us realize that we are collectively far safer if all cars were being driven by computers—particularly in this age when it seems like every second person on the highway is updating their social media status while trying to pass a truck. Have you ever considered that the major task of a pilot on a commercial airliner is to greet you when you deplane to reassure you that there was a live human body in control—even though the plane was likely on autopilot for the vast majority of the trip. (Ever notice how frequently the pilot does not greet you as you disembark after a particularly rough landing in calm weather—for which landing is the one time that the pilot is in control.)

Risk is about people. However risk is not necessarily about people behaving badly, or stupidly or for that matter even consistently brilliantly. People are people and we all do things that make us wince with hindsight. We all have regrets, and we all have moments of triumph. The problem is that we cannot control those moments as much as we would like to. We get tired, or lazy or perhaps brilliantly inspired. We know some things, we don't know other things, and we guess at things way more than we would honestly estimate that we do. We also have very different world views and experiences that color our world views and how we interpret things.

Risk and people arise in two forms that are very different and which pose different challenges and different management tasks for the risk manager. People acting individually or in a small group create risk, and people acting as part of a larger group or a sociological system create risks. Risk management to the individual or the small group is very different than risk management of the larger sociological group. Risk management of the individual involves idiosyncratic risks that are likely to appear to be more random in nature. Group risks on the contrary are likely to demonstrate some form of complex emergence. In other words, a risk evolving from the interactions of a group of people is likely to show patterns (at least with hindsight), but those patterns will be random, unpredictable, and not amenable to management in the conventional sense. To understand and manage the risk of individual actions, you need empathy. To understand and manage the risks of the sociological group, the risk manager needs to develop a sociological imagination.

Empathy and Sociological Imagination

Empathy and sociological imagination are the two necessary abilities needed for the risk management of people. Unfortunately empathy and having a sociological imagination are things that we as humans suck at. This presents a real risk problem, as empathy and a sociological imagination are two vital keys to successful risk management. Another difficulty is that although we, as humans, suck at empathy and having a sociological imagination, a computer sucks at it even more, and additionally, a rule-based series of risk management processes sucks even more!

Empathy is understanding the emotions and opinions of another person or group of persons. It is not necessarily agreeing with those emotions or opinions, nor is it necessarily having sympathy for those emotions or opinions. Empathy and sympathy are two different things. While it is nice to have sympathy, sympathy does not help that much in decision making; empathy does in that it allows you to understand how to frame an effective solution in the context of how it will be interpreted by the person(s) it is geared toward.

Harvard Business School professor Rita McGrath says that business is now entering a new era of empathy.[2] In a Harvard Business Review article, Professor McGrath states that the first era of management was the era of simply figuring out how to do things. This was the case during the industrial revolution when how to harness the power of the steam

---

[2] McGrath, R., Management's Three Eras: A Brief History, Harvard Business Review, July 2014.

engine, how to engineer things, and how to efficiently manufacture products were key. This led to the era of Scientific Management that brought forth Taylorism and the assembly line and led to the creation of MBA programs and business as a field of academic study. However, McGrath argues that we are entering into a third age: the Age of Empathy. The Age of Empathy involves not knowledge but instead requires an innate understanding of how individuals, and perhaps more importantly a collection of individuals, behave and change in their behaviors depending on context.

What many risk managers and technocrats often fail to understand is that the actions of individuals and collections of individuals cannot be reduced to a set of consistent and rational principles. Humans are unique and their thoughts, dreams, and actions cannot be reduced to a formula or a set of principles. Referring back to Systems Theory from Chapter 3, the Age of Empathy requires a complexity mindset rather than complicated knowledge. It is an important point when remembering that people and their actions need to be at the core of risk management.

Related to empathy is having a sociological imagination. If empathy could be described as understanding the emotions of an individual, then a sociological imagination is the group equivalent. Sociological imagination is thinking outside of your own worldview of things and seeing them as a broader swath of society might imagine them or interpret them. The sociologist C. Wright Mills coined the term *sociological imagination* to point out the importance of not only understanding how we as individuals think and act, but how we as part of a society think and act. Complexity, and its associated phenomena of emergence demonstrates that the collective outcome of a social group is not the simple summation of the actions of the individuals that comprise the group.

Organizations are part of a sociological group. Industries are also a sociological group, and in fact the entire economy can be properly thought of as a sociological group. If managing with empathy requires a complexity mindset, managing with a sociological imagination demands it. Managing with a sociological mindset means that the risk manager is considering the possible set of connections and adaptive behaviors that the people in the group may form and the emergent patterns and outcomes that may come from the group.

If one was to typecast risk managers, they would generally not be considered to be too high on the scale for either empathy or for having sociological imaginations. Risk managers as a group would more typically be typecast as being technocrats, as individuals tied to careful rational thought. Basically risk managers and the risk strategies they promote and implement would probably be most accurately typecast as the antithesis of empathy and a sociological imagination. This fact may be the root cause of a significant amount of risk in organizations—namely the disconnect between risk management systems and the people who are the source of the risk.

The Importance of Design

How important is design in your risk management system? How many designers or ergonomic engineers are in your organization's risk management group? If you accept that risk management is largely caused by people, shouldn't the risk management function and processes be designed for people. Shouldn't the risk management functions be designed with the psychological needs, the sociological needs, and the ergonomics needs of the implementers and the users and the beneficiaries of the risk management in mind?

Design is thought of as a nice to have, not a must have. Design is too often thought of in terms of aesthetics, rather than functionality and efficiency. Good design can help people not only conduct actions that are more effective but also follow the rules more than they would without good design.

The principles of good design are beyond the scope of this book (and beyond my area of knowledge), but it should be obvious that good design is central to good risk management.

Delphi Method

A very powerful method for uncovering potential risks and designing effective risk management is the Delphi Method. The Delphi Method is a discussion technique that allows for cognitive diversity to produce innovative ideas to come to the fore without the concerns of groupthink and bureaucracy preventing progress.

In the Delphi Method, a diverse group of individuals is given the task of risk identification as well as risk ranking. The better the diversity, the better the results are likely to be. The group does not (in fact should not) be composed of risk experts, but instead it should be composed of a sample of individuals associated with the organization from different divisions of the organization, from different stakeholder groups, such as suppliers or customers, and from different levels ranging from frontline employees through to senior managers. The group is then given a question such as what are the major risks associated with a given task or a given unit of the organization. (It should be pointed out to the assembled group that the definition of risk is that of the possibility that bad or good things may happen.) Individuals in the group are asked to anonymously list what they believe are the primary risks. A facilitator will then list the risks and facilitate a discussion about the listed risks. Then the group is asked to rank the risks anonymously. The results of the ranking (done by anonymous voting) are then presented to the group and another facilitated discussion takes place. The process of discussion and anonymous voting is continued until a consensus ranking is achieved.

The diverse individuals (again, they are not risk experts and come from a wide range of functions and levels of experience and seniority) and anonymous voting are the two keys to a successful outcome for the Delphi Method. The diversity of individuals brings out unique ideas that would likely not be thought of if a group of experts well acquainted with the situation had been assembled. Admittedly many of the risks or ideas put forward will not be valid or appropriate, but they will be quickly dismissed by the process. The anonymous voting prevents group think and it also prevents a dominant member of the group being able to sway the group. Finally it avoids the Asch effect that is so prevalent when groups or town halls are convened to assess a situation or suggest solutions.

The Asch effect comes from the research of Solomon Asch. In a brilliantly designed experiment, Asch had groups of experimental subjects tell him which of a group of drawn lines was the longest. Everyone in each group observing the lines was in on the experiment except for a single individual. In other words, each group was composed of actors, while there was only one true experimental subject who was responding with their own nonprearranged answer. The group would be shown a set of lines labeled A, B, C, and D. One of the lines would be obviously much longer than the others—say for instance line C might be drawn much longer than lines A, B, and D. Each of the experimental subjects who were actors would, however respond that line B was the longest line. This was obviously not true, but when the true experimental subject was asked to state in front of the others what line they thought was the longest they would also respond with line B. The purpose of the experiment was to show that our desire to conform to the thoughts of a group, even though the group was obviously wrong, was greater than our desire to state what we truly believed to be true.

The Delphi Method can also be used as a way to identify a way to manage risk. Once the risks have been identified and ranked, the group (or even a different group) can then be asked the question of how to manage the most important risks. The Delphi Method can then be repeated to rank the most effective risk management strategies. This naturally provides buy-in to the risk management response as it provides solutions that are coming from a diverse group rather than from a specific unit of the organization or solely from the senior management team. If you are part of the group that develops a suggestion, you will be much more vested in seeing that suggestion be successful.

In sum, the Delphi Method provides a very effective way for the organization to learn about its risks in a way that avoids relying on "the usual suspects." It does so in a way that provides unique ideas, avoids groupthink, and allows for the measurement (or at least a ranking) of nonquantitative risks. Additionally, the Delphi Method creates buy in as it is a risk identification that involves the individuals who have to implement the risk management plan and have a vested role in seeing their plan be successful.

Concluding Thoughts

All too often, risk is thought to be caused by something outside of our control. Risk is credited to freak occurrences, failure of materials, failure of processes, random acts of God, and so on. All of these explanations are random or have their basis in complicated systems. Admittedly these things do cause risk events, but these explanations are rarely the root causes. People either individually, or as a group are much more frequently the root cause. Additionally as people are complex, we can also add complexity as a frequent root cause of risk. This creates a disconnect between how we think about risk and what causes risk. Risk is generally not totally random, and likewise risk is generally not complicated. Risk is people and complexity.

Dealing with people and complexity requires a different way of thinking, a different mindset from the complicated calculation of risk metrics. It requires thinking in terms of complexity, empathy, and having a sociological imagination. Admittedly easier said than done, but some risk managers have the ability to think that way. Training risk managers does not solely require mastering mathematics, or learning engineering principles, but also requires mastering people and why people act the way they do in given situations.

There is one final principle to remember about people. You can't fix stupid, and people (including me, and perhaps especially me) sometimes act stupidly. This is where design kicks in to make the risk system as robust to failure as possible. But it is also why the old risk adage that "the only perfect hedge is in a Japanese Garden" is so true.

CHAPTER 5

Are Risk Frameworks Evil?

There is a wide variety of books, seminars, and consultants who stress that if the proper framework for risk management is developed, then the rest of the task of creating an effective risk management function will fall into place. I am not one of those consultants, and this is not one of those risk books. After the twin evils of regulation for the sake of appearing to do something, and the overmathmatization of risk management based on the false belief that risk is complicated, I believe that risk frameworks are one of the worst things to happen to risk management. Instead of promoting a specific risk framework, or even risk frameworks in general, this chapter puts forward and defends the bold argument that risk frameworks are evil!

Frameworks are ideal for houses and other structures where stability and inflexibility are needed, but risk management is not one of those structures. By its nature, risk management needs to be flexible and adaptable. While conceptually risk frameworks are good, the reality is that they are almost always taken far too seriously (almost religiously) by organizations, and what is left is a rigid shell that is best left for other, yet unknown purposes.

As discussed in the previous chapter, What Causes Risk, risk is fundamentally caused by people and complexity. The rest of risk management is relatively straightforward. People and complexity, however, do not fit well into frameworks, as least not the creative entrepreneurial types who are so key to success in this age of knowledge-based organizations. Frameworks are far better suited in simple or complicated, rather than complex systems. In an earlier time and place, when a human operating the machine was the basis of competitive advantage, rigid frameworks for both operations and risk management had a definite role to play. Whether we like it or not though, those days have either long past or are on their way out the door, so some new thinking is needed.

Two Popular Frameworks

The two main risk frameworks utilized by organizations are the Committee of Sponsoring Organizations of the Treadway Commission (more commonly known as COSO)[1] and the International Organization for Standardization's ISO 31000.[2] Both frameworks share a lot in common, although they use somewhat different structures and their language is somewhat different. The common elements begin by acknowledging that the objective of the risk management must be clearly articulated. Other common elements include risk identification, risk assessment, risk monitoring and communication, and importantly, the realization that risk management is an ongoing and continuous process.

Both the COSO and ISO frameworks, as well as the various frameworks put forward by the various consultants engaged in risk management, have attractive features. For organizations that are new to risk management, they provide a quick-start method for getting started in risk management. They also get a robust discussion started on risk management. Indeed, they are likely to start a more thorough discussion on risk management than many organizations may be ready for. The thoroughness of the frameworks is likely to scare, or even shock, organizations into action as an organization compares what it is doing in risk management compared with the comprehensive frameworks that exist.

Both the COSO and ISO 31000 frameworks have been designed by committees. We are all aware of the organizational jokes about the effectiveness of committees (e.g., a camel looks like a horse designed by a committee). They are frameworks that have been designed to be as widely applicable as possible, but risk is almost always specific to a given organization and to the uniqueness of an organization's culture. It frequently becomes more work to adopting a given risk framework to an organization than it would be to develop a unique framework from scratch, which is tailored for the organization. My experience has been that adopting an existing framework for an organization is akin to adapting a square peg for a round hole.

---

[1] http://www.coso.org/.

[2] http://www.iso.org/iso/home/standards/iso31000.htm.

When an organization first adopts a risk framework, it is undoubtedly a positive for risk management. Just as a house needs a good framework, some form of structure for risk management is generally good. However, the construction of a house does not stop with the framework. Also, one does not generally define their house by its framework. A framework is not what changes a house into a home—to adopt a common phrase for a new use. A home is defined by its decorating and the people who live within its walls that hide the framework. A home is even better defined by the memories, both good and bad, that occurred while the occupants lived there. Likewise, the risk framework in an organization should be similarly hidden. What should define risk management is the culture of the people in the organization and the intuitive wisdom—both good and bad—that they have built up through their collective experiences.

A central issue is that a risk framework can very quickly become risk management of an organization with little in the way of compromises to account for the distinctiveness of the adopting organization and its culture. It is the equivalent of the framing of a house becoming the house. That of course would be silly. Thus, while risk frameworks help to start defining the risk management function, one has to quickly ask whether the risk frameworks are better for the organizations that they are supposedly supposed to help or are they better for the consulting organizations?

Risk frameworks by their nature stifle many of the appropriate risk activities while encouraging many negative activities. Risk frameworks stifle creativity, independent thinking, value differentiation, responsibility, timeliness, opportunities, black swans, and the use of risk management as a strategic tool for competitive advantage. Additionally, risk frameworks encourage laziness, box-ticking, bureaucracy, costs, inefficiency, hopelessness, confusion between auditing and risk management, and finally, act as a catalyst for risk homeostasis.

If this is true, then why have risk frameworks become so popular for risk management? In my opinion, risk frameworks have become popular for two reasons; they give the appearance of doing something concrete, and they allow consultants to firmly embed themselves within a client in order to secure longer consulting contracts, and thus, larger consulting fees.

Risk frameworks outline a process that provides an organization a false level of comfort that they are doing something concrete about risk management. They allow for the quick and easy development of lots of charts, flow diagrams, and progress reports. These elements, in turn, give a level of comfort to senior management, the board, and regulators that the firm is actively doing stuff. It also lends credibility to the risk management department, as they can shows lots of impressive-looking diagrams and schematics. Perhaps, the greatest application of this is for consultants who can show a comprehensive framework that impresses clients and potential clients that they, the consultants, truly have the key for completely solving an organization's risk management issues.

As will be discussed more fully later in this chapter, risk frameworks enable and encourage bureaucracy. Thus, risk frameworks are a wonderful tool for consultants to use to increase their billable hours. The more comprehensive and the more integrated the framework, the more necessary it is to have consultants manage it. Also, each consultant organization has their proprietary framework, which binds it to an organization. It is kind of like buying a specific brand of shaver—once you buy a shaver, you are more or less committed to using that brand of razor for a significant period of time, and thus, repeatedly buying replacement blades from that same brand. Adopting a consulting firm's "proprietary" framework also implies you will be stuck with that consulting firm for a while, as the entire organization will likely let out a collective groan at the thought of switching to a new consulting firm's "proprietary" framework.

Risk frameworks do have their place in an organization's risk toolbox. However, it is important that an organization thinks carefully about the pros and cons of adapting a specific framework, and how they plan on using a framework. Very quickly, frameworks tend to become risk management, rather than a guide for risk management. In other words, people manage to the framework, rather than manage risk—two very different things. Risk frameworks can very quickly turn from something that is useful to something that is evil. Frameworks have risk elements that they stifle and elements that they enable. It is critical that a firm becomes aware of these elements.

Stifling Abilities of Frameworks

The first thing that risk frameworks stifle is creativity. Creativity is absolutely key for effective risk management. I claim that the mere fact that you acknowledge that a risk exists automatically increases the probability and magnitude of it occurring if it is a good risk, while it also simultaneously decreases the probability and severity of it occurring if it is a bad risk. Being creative helps the risk management team and the organization to see risks—both good and bad—while there is still time to manage them.

Risk management frameworks are great at cataloguing existing risks and historical risks, but by their nature, they are very poor at creatively seeing new risks on the horizon. Risk frameworks tend to be comprehensive. This is a plus for an organization brand-new to risk management. However, it tends to be an even bigger plus for the risk management organizations that provide an army of consultants to implement and maintain the framework.

The fact that the frameworks are so comprehensive is very much a two-edged sword. The extensive development of the frameworks means that a lot of experience has gone into their creation. This helps an organization uncover risks and elements of risks that they likely would have overlooked. But the other side of the sword is that the comprehensiveness can easily overwhelm an organization. The second drawback is, with comprehensiveness, comes a complacency that everything has already been incorporated into the framework. The forest often gets lost for the trees in the effort to be true to the comprehensiveness of the framework.

Sometimes, the best way to generate creativity is to start with a blank sheet of paper. An existing framework is anything but a blank sheet of paper. Too conveniently, it has a label and a place for seemingly everything, and it is all too easy to quickly develop faith that the framework will cover everything that needs to be covered. This feeling of more than adequate coverage quashes the questions that lead to creativity.

In high school, I worked for a well-known sales organization selling vacuum cleaners. It was to earn extra money for all those things that a young person thinks they need while going through high school. Before setting out on our nightly sales calls, there would be an office meeting, basically a sales pep rally. One night, relatively soon after I started working, one of the newer sales agents queried whether or not there might be a better way to conduct operations. I remember the office manager walked over to the eager, but unsuspecting newbie and letting loose with a tirade about how the company had been in business for decades and what in blazes could this relatively inexperienced sales agent possibly know that the company had not already thought of. The newbie left the company that night in tears, and I considered the statement "what makes you think you are smarter than the organization?" to be one of the dumbest management sayings ever. I quit two nights later, and a couple of months later, the organization went bankrupt.

A comprehensive framework makes everyone think they are "dumber than the framework," which is false, demoralizing, stupid, and obnoxious.

A comprehensive framework creates the illusion that everything is covered. However, there is value in feeling exposed. A feeling of exposure creates a valuable paranoia. Intel CEO Andy Grove famously said that only the paranoid survive, and it is likely that he is correct in thinking so. When one is paranoid and feeling exposed, it leads one to be more aware. It leads one to either find ways to eliminate the exposure or exploit the exposure. These are instincts that should not be stifled.

Related to stifling creativity, risk frameworks stifle thinking. Having an established framework firstly means that one does not have to think hard about how to create it. In fact, the realization that a framework already exists means one does not need to think at all in order to create it or to maintain it. While it is obviously convenient and a real time and energy saver, not having done the hard work of designing a risk framework means that one likely does not know both the strengths and weaknesses of the framework.

It is simply too easy to accept a new framework without thinking. Sure there is likely to be push back at inception as various groups with special interests in the status quo will attack the framework pointing out its weaknesses and inconsistencies. Likely almost everyone in the organization will take a shot at knocking down the proposed framework, but will ultimately accept it, as intuitively they know it is much easier to criticize than it is to create a better solution—although many likely better ideas exist.

Without the hard thinking in creating a framework, there will also not be the same level of ownership. This is particularly so if a consulting team is engaged to implement the framework. In such cases, the "intelligence" of the framework will be seen to exist with the consultants, and not the organization. There will be a lack of internal ownership. This has implications not only for ownership, but also for future adaptation. Who wants to fix another's mistakes? If the risk management function is built internally, then the intellectual energy that created it will also have a vested interest in maintaining it and keeping it current. Engaged thinking minds are always a good thing.

Perhaps, the biggest lack of thinking about risk is that there is the potential for risk management to become management to the framework. A framework does not think. A framework does not adapt. A framework is static, and not dynamic. Management to such a framework is, thus, anything but effective risk management.

No matter how good a risk framework is, there will be problems. No risk framework can pick up and help to exploit every type of opportunity, and likewise, there will be mistakes. With a standard framework, however, there is no ownership of the framework, and with no ownership, there is no responsibility or accountability.

Risk frameworks create the illusion that risk management is the role and responsibility of the framework. If something goes wrong, it is the framework's fault, and not the fault of the employee or the department or even the organization. The framework becomes a very easy and convenient scapegoat. This is not productive and will only cause problems down the road.

Using an existing risk management framework is akin to inviting all of your friends over for a summer barbeque and serving them fast food hamburgers. Simply put, there is an embarrassing lack of value differentiation, and the next time you throw a summer time party, it is likely that your friends will be "busy" with some other activity. Unless the risk framework allows the organization to differentiate its risk function, to show its creative abilities and how it can support, rather than hinder the organization's activities, then the risk function will not be seen to be a competitive value-adding function. A rigid risk framework stifles out-of-the-box thinking, and thus, destroys the potential for unique value-adding functionality. This is particularly true if one of the standard risk frameworks is utilized.

Risk management should be a strategic tool for competitive advantage. In turn, a competitive advantage needs to be dynamic, not static.

Risk frameworks by comparison tend to become bureaucratic monstrosities that take a life of their own. Risk management becomes a function that is done for the sake of the risk framework, rather than a function for risk management and competitive advantage.

Perhaps most troubling is that risk frameworks stifle having a complexity mindset. By definition, frameworks are rigid. Complexity is anything but rigid. Frameworks encourage processes and categorization, while complexity and emergence defy replicating processes and categorization. Frameworks are constructed for a solve mentality, while complexity requires a *mange, not solve* attitude. In essence, it is difficult to manage with a complexity mindset when the order and structure of a framework are being imposed.

Enabling Activities of Risk Frameworks

Risk frameworks are not only stifling, but they are also great enablers. To begin, risk frameworks enable laziness. One does not need to be engaged in thinking about risk if a comprehensive risk framework is in place, do they? As long as the framework is in place, all one feels that they need to do is follow the framework. This is an incredibly lazy way to engage in risk management. It also enables blind spots to risk to develop and grow. If one is following the framework, they feel safe, but no framework is comprehensive enough to cover all possible risks. It means that new or evolving risks are missed. Problems arise unnoticed, and opportunities are missed.

Related to laziness in risk thinking is the lack of accountability. With a prominent risk framework in place, if there is a problem in risk management, or a risk inefficiency, then everyone assumes it must be the problem of the framework. A strong risk framework can very quickly become an enabler for the shedding of accountability. Risk laziness and risk accountability are closely linked. If one believes that they play second fiddle to the risk framework, they are just going to follow the first fiddle—namely just follow the framework. In doing so, they automatically abdicate responsibility for risk, and with this abdication, it is extremely tempting to become lazy in one's risk thinking, risk awareness, and risk creativity.

In a similar vein, risk frameworks are great enablers of risk management by box-ticking. Management to the framework almost always leads to managing to a process. The question of whether risk management should be process-based or judgment-based is covered in Chapter 7, but for now, it suffices to say that managing to a process implies that one is focusing more on complicated aspects of risk and misses those situations that are complex and require judgment. Additionally, the process becomes the end, rather than the means to an end.

Risk management by risk framework box-ticking leads to, perhaps, the two biggest enabling issues with risk frameworks; they encourage a focus on auditing, rather than risk management, and they encourage bureaucracy, rather than lean and mean efficiency.

In many organizations, auditing and risk management are synonymous. That is, a mistake; a big mistake. Auditing is checking that boxes have been ticked. It is a form of quality control. It says nothing about whether or not the right boxes are being ticked, and it adds little to nothing about what the definition of quality is. Auditing is an important task, and a necessary task, but it is not risk management. For starters, auditing is a passive task, while risk management is an active task. Auditing focuses on what was, and to a lesser extent on what is, while risk management is focused on the future. Auditing records; risk management creates. Auditing and risk management are totally different tasks and require totally different mindsets. They should never be confused for each other, and they should never be commingled.

John Fraser, former Chief Risk Officer at Hydro One, is a risk manager who many (including me) consider to be the Warren Buffet of enterprise risk management; he is a true leader in the field, who thinks very clearly and intelligently about risk management issues. While at Hydro One, he was also the firm's Chief Auditing Officer. When one went into Mr. Fraser's office, one saw two caps prominently displayed; one labeled Chief Risk Officer, and one labeled Chief Auditing Officer. John Fraser had the two caps to remind himself that the two tasks are very different and that one looks ridiculous if one tries to wear the two hats at the same time. Confusing or commingling auditing and risk management implies that one does not understand nor appreciate the importance of either of the tasks.

Risk frameworks very quickly become "the process," and a "process" quickly becomes a bureaucracy. Most of the popular risk frameworks were designed to be as general as possible and as widely applicable as possible. Thus, they tend to be much more cumbersome than necessary for most organizations. Add to that the fact already discussed that risk frameworks are ideal for creating long-lasting work for consultants, and one can see how risk frameworks become enablers of fat bureaucracies. Without any prioritization of parts of the framework, every part of the framework becomes mission-critical, which of course means that no part becomes mission-critical. What it does mean that each part of the framework develops its own bureaucracy which limits the integration, which was one of the main points of having a framework in the first place.

One of the most popular frameworks, the COSO framework, has a "cube" as its structure.[3] On the front face of the cube are eight elements that a company should undertake for risk management: (1) internal environment, (2) objective setting, (3) event identification, (4) risk assessment, (5) risk response, (6) control activities, (7) information and communication, and (8) monitoring. Along the top face of the cube are four functions that an organization should conduct the eight elements on. They are: (1) strategic, (2) operations, (3) reporting, and (4) compliance. Finally, on the side face of the cube are four levels of the organization at which the elements for each of the functions must be conducted. These four levels are: (1) subsidiary, (2) business unit, (3) division, and (4) entity level. In total, there are 128 "boxes" that make up the "cube." It quickly becomes obvious how such a framework becomes a bloated bureaucracy that can easily take on a life of its own.

My first exposure to the COSO cube was at a workshop conducted by a consulting group. The four-day workshop consisted of the facilitator giving a detailed analysis of each of the 128 boxes and a laundry list of details to be managed for each box. Seminar participants debated the nuances of managing one box versus another box. It was completely

---

[3] http://www.coso.org/documents/coso_erm_executivesummary.pdf. As this chapter is being written, a new, more simplified COSO framework is being developed and discussed.

nauseating. No wonder organizations outsource their risk management to consultants!

While risk management needs to be thorough, it also needs to be practical. If something is so comprehensive and energy-consuming in its application that it sucks the life out of the rest of the organization, then it is obvious that it is overkill.

I believe that risk management is most effective when it has structure for the simple and complicated parts, and for everything else, very limited structure to enable the flexibility needed for dealing with complexity. I also believe that risk management should be as lean as possible. Leanness implies that everyone is responsible for risk. If the risk function is seen as a fully staffed and resourced standalone bureaucracy, then there is a high likelihood that risk management becomes an end to itself, and not a means for organizational success. If everyone is aware that they need to be managing risk because there is no special unit focused on it, then the overall focus on risk becomes much greater. Risk frameworks, however, by their nature, discourage leanness and promote and enable bureaucracy.

Ultimately, a strong risk framework actually decreases the risk effectiveness of an organization. Effectively, risk frameworks bring on risk homeostasis, the topic of Chapter 9. In essence, risk homeostasis means that if the firm has a prominent risk framework in place, then it is counterintuitively likely to be worse off in its risk management.

Concluding Thoughts

Risk frameworks have their place. However, risk frameworks have aspects that stifle those characteristics that are inherent in good risk management and enable characteristics that are bad for effective risk management. Risk frameworks are like management by recipe. As discussed in Chapter 3, recipes and checklists are fine when issues are simple in nature, but by nature, most risks tend to be complex. Complexity demands a more enlightened manager than one who is simply following a recipe.

As an organization matures in its risk thinking and in its risk management, the need for a risk framework diminishes. Ideally, I believe that many organizations would be better served by striving to wean themselves off of their risk framework and instead focusing on development of a more dynamic, in-the-moment, frameworkless mode of managing risk. While being frameworkless might seem a bit more chaotic, it is likely to be much more effective and much more efficient for risk management.

Are risk frameworks evil? Conceptually, the answer is no. Frameworks are not inherently evil, but they have become that way as they stifle many of the characteristics of positive risk management and enable other characteristics of bad risk management. Too often, a great sounding risk framework creates a risk mentality in an organization that makes it like the frog in water that is slowly heated. The frog never realizes it is being boiled to death, and likewise, an organization being taken over by a risk framework never realizes it is abandoning good risk management and replacing it with evil.

CHAPTER 6

Does Risk Management Add Value?

In the 1990s, as derivatives started to come to the fore as a risk management tool, many companies started to think of risk management as a profit center—or at least think of their financial risk management activities as having the potential for profit generation. In fact, I can think of one company that regularly put forward guidance in their earnings projections the profit they expected to generate from their hedging activities. While most companies were not that bold, nor as brazen in their expectations for profits from hedging activities, there were several surveys conducted asking companies whether they viewed risk management as a profit-generating activity or as a cost center. Then, the hedging debacles, such as those at Gibson Greetings, Procter and Gamble, and Mettalgellschaft, came along and the idea of risk management generating profits soon became a taboo concept.

Despite this rather embarrassing past, it is worth revisiting the question of if risk management adds value, and if so, how can that value be maximized. The question is not just a question of the bottom line for the company, but it is at the center of both the risk philosophy of the organization, and perhaps, more importantly, the risk culture of the firm. Also, I need to stress upfront that I am not talking about making profits from risk management through the use of financial engineering. The issue is whether risk management concepts can be used for value, rather than just being used to prevent the loss of value. So let's begin the discussion.

Revisiting the Definition of Risk

If you have read this far, then it is likely that you have not been totally revolted by the definitions of risk and of risk management that were given in Chapter 1: What Is Risk? Chapter 1 outlined the definition of risk as "the possibility that bad or good things may happen." One the central elements of this definition is that it is that risk is not just concerned about bad things, or the downside, but also about good things, the upside. This naturally leads to the definition of risk management as "managing so as to increase the probability and magnitude of good risk events occurring while simultaneously managing so as to decrease the probability of bad risk events occurring."

The issue becomes the ratio of focus on "managing so as to increase the probability and magnitude of good risk events" to the focus on "managing so as to decrease the probability and severity of bad risk events." How does that ratio look for your organization? If your organization is typical, then the focus is virtually 100 percent on the downside risk and zero on the upside risk.

When I bring this point up in seminars, it is pointed out to me that the upside is the job of the CEO and the marketing department. On one level, that is a fair enough statement, but in how many organizations does the CEO or marketing have the same expertise in dealing with probabilities and uncertainties as the risk department? How many utilize the tools of risk management? Does it mean that risk management does not have the capability to think in terms of opportunities?

I was once at a large corporate meeting in which the CEO of the organization was addressing the employees of an entire division of the company at a weekend offsite. The CEO's talk was centered around the theme of how his job was to look out for storm clouds on the horizon. In essence, it was a talk geared for risk management and essentially billed as a rah-rah speech for the importance of risk management. I was invited beforehand to be one of the participants at the conference to publically ask a question of the CEO. As the CEO talked about all of the bad things and challenges he foresaw for the organization, all I could think about was what a gloomy talk. Who at a supposedly uplifting corporate offsite wants to hear about how all the CEO thinks about is storm clouds? Kind of takes the shine off of what should otherwise be an inspiring weekend retreat. When it came time for me to ask my question, I decided to try to ask a question that might improve the mood and the tone in the room. The question I asked was, "My mother always told me that for every cloud there is a silver lining. Would you be willing to share with us some of the silver linings that you see in your gazing at clouds on the horizon?" The response I got was a glare of contempt, a very brief, vague, and mumbled answer, and a reaction of incredulousness that I would ask such an inappropriate and stupid question. To this day, I still think it was a good question and a valid question, but it also struck home how ingrained pessimism is in the mindsets of risk managers. It is time to change and expand that ingrained perception.

Shut Down Risk as the "Department of No"

There are many reasons to re-examine the question of whether (and how) the risk management function can add value. A primary reason (at least in my mind) is to shut down risk as the "Department of No," and the negative attitude and aura around risk that this creates. As management consultant and sales expert Zig Ziglar clearly stated, "positive thinking will let you do everything better than negative thinking will." This holds not only for sales management, but for risk management as well.

I will discuss much more about having a positive attitude around risk in Chapter 8 on Risk Culture, but for the current question of risk adding value, it is important to realize the importance for the attitude toward the risk management function in having both positive risk as well as negative risk to be managed.

In many of the management seminars on risk management that I conduct, I ask the question of if it is reasonable to label the risk department as the "Department of No"? The question is always met with a series of uncomfortable smiles and a resounding answer of yes—it is reasonable to label risk as the "Department of No." Because of this reputation, risk is seen as a hostile internal enemy, rather than a function that is helping the organization achieve new and greater goals. Risk is a drag, not a catalyst. That is sad, unfortunate, and in my opinion, just plain stupid.

As the "Department of No," risk is automatically seen as something to be avoided, and with it the services and the ideas of risk as also something to be avoided. With this attitude, risk processes, risk concepts, and risk intelligence are something that are seen as an obstacle, instead of allies or as helpful tools that can be utilized. Risk is avoided and distained. The risk department is seen as a unit that the rest of the organization has to constantly go to battle against. If not battle, at least a unit that will put lots of roadblocks, red tape, and time-wasting bureaucratic analysis on the path to success.

When risk is seen as the "Department of No," it implies that the risk management function is operating at 50 percent efficiency at best. It is operating to prevent things from happening, rather than to help things happen. It is time to change that and make risk the "Department of How Can We Do It Better," or "How Can We Do It More Risk Intelligently."

Risk as "The Department of How to Do It Better"

Setting up the risk management function as "The Department of How to Do It Better" is the first step to making risk a value-adding function with a wide variety of side benefits.

Risk management, of all forms, borrows heavily from engineering. Engineering is focused on how to make things happen, how to build things, how to invent things. Yes, a large part of making things happen is preventing bad things from happening, but the focus is on making things, on making progress, on getting things built, and on improving things. Advances in engineering are focused on improvement, not on prevention.

With risk as the "Department of How to Do Things Better," risk becomes the ally. Instead of something to be avoided, it is a function to be brought on at the earliest stages of an idea or a project. Risk becomes a partner in success, rather than a drag. Risk starts to work not at 50 percent efficiency, but much closer to 100 percent efficiency, with the focus on managing both upside as well as downside risks.

With this change in philosophy, the responses to risk become more complete. The most value-added addition to the risk responses is that risk may not only be something to be avoided, but something that in certain situations that should be embraced. The power of using the risk management function as a group that allows risk to be embraced is something that is extremely powerful.

With risk as the "Department of How to Do It Better," there is likely to be an ironic side effect. Seen as an enabler, the risk management function will be brought into the design process much earlier in the development of new products. This fact by itself will enhance the design process, making it more attuned to not only the upside as well as the downside—much more than it would be with risk being avoided until necessary. With risk principles embedded in the design from the get-go, the ultimate design is likely to be much more efficient at managing the downside than if risk is incorporated at the final stages. By focusing on the upside, the downside becomes more effectively managed—a most pleasant, ironic, and unintended consequence of the change in attitude.

If risk is brought in as an ally, the role of design in risk management comes more to the forefront. This raises the interesting question of the role of design in risk. How many designers are part of your risk management function? Is design important in your risk management function? Does design matter? Or, is your risk management function focused almost exclusively on metrics?

Risk is a social activity. As discussed in Chapter 4, the main cause of preventable risk is people; people acting on their own and people acting as part of a society either as part of a large or a small group or even as part of a much larger sociological phenomenon. Yet, my experience is that few risk departments have designers or sociologists as part of their staff, and thus, have little or no appreciation of the power of design or sociological thinking.

Design in risk management should extend from the design of processes, to the design of training to the design of the risk metrics that the firm utilizes.

Risk Metrics for Value

Returning to Peter Drucker's famous adage of "what gets measured gets managed," a quick examination of the most popular risk metrics shows the bias toward negativity. While probably the most popular quantitative measure for risk, standard deviation (or equivalently the variance), does measure the deviations both upside and downside from the average, it is the focus on the downside that predominates.

The primary evidence that the downside dominates when using standard deviation is that value at risk, or commonly known as VAR, is a measure of the most severe downside possibilities. VAR measures the negative outcome that is probable given a stated level of confidence. For instance, a VAR of 10 million dollars with a 5 percent confidence level over 10 days means that the firm can be expected to have a loss of 10 million or more in any given 10-day period at least 5 percent of the time. There are a lot of well-known problems and drawbacks and issues with using VAR as a risk measure, but for our purposes, here the issue is that it brings the whole focus of risk management on the downside. Rarely, if ever, is upside VAR discussed or even considered although such a measure could be extremely useful.

Even if management uses standard deviation as its measure without specifically picking out the downside, the most popular application of standard deviation is to make it as small as possible. It is a general rule of thumb that the smaller the standard deviation of possibilities, the better. However, this ignores that a larger standard deviation also implies the potential upside is greater.

One way around this issue, and a measure to bring the positive aspects of upside risk into focus, is to use semi-standard deviation (or equivalently semivariance). Semi-standard deviation separately breaks the measurement of risk into upside risk and downside risk. To understand semi-standard deviation, consider the following three formulas. The first formula is the well-known formula for standard deviation. The following two formulas are for positive semi-standard deviation and for negative semi-standard deviation:

$$\text{Standard Deviation} = \left(\sum_{i=1}^{n} \frac{\left(x_i - x_{avg}\right)^2}{n-1} \right)^{\frac{1}{2}}$$

$$\text{Positive Semi-Standard Deviation} = \left(\sum_{i=1, iff\ x_i > x_{avg}}^{n} \frac{\left(x_i - x_{avg}\right)^2}{n-1} \right)^{\frac{1}{2}}$$

$$\text{Negative Semi-Standard Deviation} = \left(\sum_{i=1, iff\ x_i < x_{avg}}^{n} \frac{\left(x_i - x_{avg}\right)^2}{n-1} \right)^{\frac{1}{2}}$$

In positive semi-standard deviation, only those values where the observation, the x_i, are above the average are included, while conversely with negative semi-standard deviation only those values where x_i is less than the average are included. This breaks the deviations from the mean up into those observations that are above average, and those observations that are below average. In essence, it creates a separate measure of upside risk and another separate measure for downside risk. Thus, for the range of possible outcomes, the upside risk can be measured relative to the downside risk. It seems basic, but having such a slight and obvious change in a risk metric can significantly change the risk conversation.

A further variation is to base the upside risk and downside not relative to the mean, but instead relative to a benchmark. For instance, in the previous formulas for semi-standard deviation, the x_{avg} can be replaced with a more appropriate benchmark such as a profit goal, or an efficiency goal. This makes the relative measure of upside and downside even more relevant, as, generally, the target is not the average outcome, but some other level of performance. The conventional standard deviation formula cannot adapt in this manner to a focus relative to a benchmark, rather than relative to an average.

Deviations from a goal or benchmark are obviously more important than deviations from an average. The use of the average for the basis is, of course, a mathematical convenience that allows the well-known statistical properties of the normal distribution to be used. With semivariance, we do not have access to the established probabilities from the normal distribution and the omnipresent normal distribution tables inside the back cover of our statistics and risk textbooks. Of course, in Chapter 2, I have already discussed the presence of leptokurtosis in risk data, which frequently renders the normal distribution as an inaccurate statistical tool that gives false and misleading answers.

For the purposes of most risk management applications, the use of semi-standard deviation is much more applicable, and it has the huge advantage that it explicitly allows for the calculation of both upside and downside risk. The desire to use the simpler normal distribution and conventional standard deviation is quite understandable historically, but with the ubiquitous presence of computers, the reliance on the ease of use of the normal distribution and standard deviation is no longer justifiable.

Risk Mapping

Not all risks of course can be quantitatively measured. For some risks, the best that we can hope for is to rank them, or put them on some sort of elementary scale. Using their own judgment, or perhaps a process like the Delphi Method, it is common to create risk maps as shown in the following diagram. A risk map plots the risks of an organization on a chart that shows the probability of the risk occurring on the vertical axis, while the magnitude of the risk is plotted on the horizontal axis.

A risk map is an effective way to visualize the risks of the firm and to provide a focus on the risks that the firm should concentrate their efforts on. Specific risks can be tracked on a risk map through time. The "path" of a risk on a risk map provides one tool that an organization can use to ask how effective their risk management efforts are. Of course, the applicability of a risk map is only as good as the analysis used to determine the probability and magnitude of each risk that is mapped, and also the selection of risks that the firm chooses to map. A firm can do a wonderful job of tracking and managing risks, but if it is tracking the wrong risks, or is unaware of, or ignorant of other important risks, then the risk map and progress of risks on the risk map is a moot point.

Risk maps, however, have a built-in bias to them. For most organizations, the risk map assumes that risks are negative. Thus, the goal is to implement actions that "move" the risk from the top left-hand corner of the graph (where it has high probability and high "negative" magnitude) to the lower-right corner, (where it has lower probability and a smaller "negative" magnitude).[1] The direction in which to move risks in such a graph is shown by the arrow in the following figure:

---

[1] On a risk map, it is common to draw a circle for each risk that is plotted. The size of the circle is related to the amount of control that the firm believes it has over a risk. A small circle may indicate a small amount of control, while a large circle indicates a large amount of control. Obviously, the goal is to move the plotting of the risk to an area of less negative severity, lower probability of occurrence, and to have greater control.

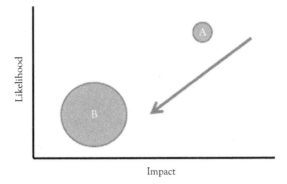

A better way to incorporate positive risk thinking is to have the X-axis explicitly have positive and negative components as illustrated in the following figure. When the risk map has the X-axis covering both negative and positive risks, then the goal is to lower the probability and magnitude of negative risks and to increase the probability and magnitude of positive risks. Note that this is exactly in accordance of my definition of risk management.

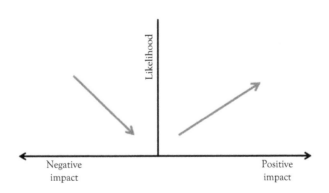

Risk management can often be rightly criticized for being too focused on the numbers. Charts, and ranking scales, while less precise than numbers, are frequently a more useful tool for many individuals who comprehend issues more fully when they can see them visually. Redefining a risk map may seem like a trivial change, but it can be a very powerful way to better incorporate the value-added potentialities of risk management.

Concluding Thoughts

My experience has been that the risk management function is massively underutilized in many organizations simply because it is seen as the "Department of No." Changing the focus to the "Department of How Can We Do It Better" or the "Department of How Can We Do It More Risk Intelligently" is a simple, but powerful move to unleash the full potential of risk management tools, tactics, and thinking.

Risk management can add value, and can add value well beyond simply preventing losses. Risk management can allow organizations to more fully exploit opportunities. The same strategies that minimize losses can be easily flipped to maximize gains.

Not only can risk management add value, I believe it must add value for those firms that want to stay competitive. Risk management is too valuable of a discipline to only utilize half of its potential.

CHAPTER 7

Should Risk Management Be Based on Process or Judgment?

We can make this chapter real short and say "it depends" to the question of whether risk management should be based on process or judgment. The problem is that, for most companies, risk management process is not set up to be quite that flexible, or that ambiguous. For a combination of reasons, mainly legal, regulatory, and insurance reasons, most companies have their risk management based almost exclusively on process, with little allowance for judgment. So much for the obvious answer of "it depends," but I guess it means that I need to continue writing some more in this chapter to explain myself.

Tomas Lopez

Let's begin our discussion with the real-life case study of Tomas Lopez. You may recall hearing about him from the news a few years ago when he had his 15 minutes of fame. Tomas Lopez was a 21-year-old lifeguard in the city of Hallandale Beach, which is in South Florida. The instructions for Tomas were very simple and very straightforward; if someone is within the roped off swimming area and in distress, go save them; if they are outside of the roped off area, call 911 and wait for them to arrive and have the 911 responders save the person. Very clear and unambiguous instructions, but for some reason, Tomas failed to follow the script. Someone who was swimming outside of the designated swimming area got into distress in the water and Tomas went into the nonroped off area and saved them. He was fired for not following the process.

If you are a risk manager, you likely believe that Tomas screwed up. He did not follow the process, and if he was not fired, then it is likely that he would have been made to attend a mandatory risk management refresher training program for his lack of risk management knowledge. If you were the person in swimming distress, then you likely have a different view on the proper action for Tomas to take. (If you were the person in swimming distress, you also have some explaining to do to account for why you were not following the rules and were swimming outside of the authorized area.)

This may appear to be a rather unusual situation—I suspect that there are few managers reading this book who are responsible for lifeguards—but the questions it raises about risk management are all too common. Ignoring the moral implications of the decision for a moment, there are excellent reasons for the policy that was in place. First, if the policy about not rescuing people outside of the roped off area is not enforced, then people will naturally enter into the water wherever they desire, which, in turn, will make the job of the lifeguard much more difficult, as they need to keep track of people over a much larger area, and thus, dramatically decrease their ability to effectively do their job. It will encourage swimming by oneself, and thus, cut down on the number of people employing the buddy system. It will also greatly increase the insurance liabilities and the potential lawsuits.

There is also the issue of what would have happened if there was an emergency with someone who was swimming in the appropriately roped off area while Tomas was saving the rule-busting swimmer. Obviously, all legal heck would have broken loose if a rule-abiding swimmer got into trouble and was not able to be saved by the diverted Tomas. (In this actual situation, other lifeguards made themselves available to cover the roped off area that Tomas abandoned.) One can easily imagine the media talking about the irresponsible lifeguard who broke procedure and saved a rule-ignoring anarchist while allowing a rule-loving patriot to perish. The letters to the editor page would have been filled with angry letters about how the fates of the two swimmers should have been reversed, and there would be follow-up stories about how Tomas had to change his identity to escape the wrath of the public that he and his family faced.

A strong argument can be made either way for the procedure that was in place for the lifeguards to follow. Having a process-based procedure helps the lifeguards—who tend to be young and without much experience in making life and death decisions—in doing their important risk management task in the best way possible. It certainly helps for legal and insurance reasons, and it allows for the critical decisions that need to be made in an instant to be deliberated at length in an atmosphere and a timeframe that is more conducive to properly thinking through all of the implications. A split-second life or death panic situation is not the best context for making decisions (or is it?).

On the other hand, the policy takes a lot of decision out of the hands of the actual employee who in the moment has the responsibility for the decision—in this case, Tomas. Some people, particularly those who want little to no responsibility whatsoever, believe that having a hard-and-fast procedure to follow is a real blessing. Indeed, some companies want employees who want that; for example, franchisers who want each of their outlets to be operated in the exact same manner. These employers do not want employees who think; they want employees who follow rules and processes unquestioningly. However, for most modern jobs, you want employees and staff who have a brain, and having a brain means people have the capability for, and a desire for, at least some accountability.

The Tomas Lopez situation also quite adequately brings home the issue of risk tolerance. I suspect few organizations have the tolerance to have their process take precedent over a preventable death. This brings up serious questions of when obedience to a process should be broken. There are clearly moral and ethical implications that go well beyond risk management.

There is also the possibility for a debate about the usefulness of Bayesian risk analysis. There is a known swimmer in distress versus the relatively low probability of a swimmer in distress in the roped off area. On the probability of saving a life, the analysis appears to be clear.

Ethical issues, insurance, risk management protocol, probability of saving a life, keeping one's job—it is a lot for a young lifeguard to have to process in real time. A well-defined and understood process makes the decision of the lifeguard much easier in the moment, but is it the right decision and what are the long-term implications? Process versus judgment is a very wicked problem.

To conclude the story of Tomas, he was fired from his lifeguarding job by the contractor who hired him, along with several other lifeguards who were involved in the rescue. The reason for the firing was that Tomas left his post to save someone in the unauthorized swimming area. The media backlash was both swift and harsh. The city and the contractor were soundly criticized, and as a result of the backlash, Tomas was offered his job back—although he refused to accept the offer to return to lifeguarding. Tomas received the key to the city for his quick decision making and was praised as a hero.

Processes and Obedience

It is likely that you think that Tomas Lopez made the right decision; or maybe you think he was wrong. I am not going to judge your opinion either way. The point of the case is to highlight the central problem behind process versus judgment. The whole reason for developing a process for risk management purposes is that you do not want any deviations from the prescribed protocol. The flip side is that it is virtually impossible to design a protocol that will cover all conceivable risk situations, and you need people to be able to use appropriate judgment for the situation at hand.

Regardless of what you believe that Tomas should have done, it is virtually certain that there will be times when you do not want the process to be followed. You might think it will be obvious when it is prudent to not follow the rules. You might believe that if such a unique and rare set of circumstances occurs that any reasonably rational employee will both know to not to follow the standard procedure, and will have the intelligence, and the will to break protocol, and will do what needs to be done. That is a lot of "ands" and a lot of wishful thinking—particularly, if the risk management function has been set up to be rules-based versus judgment-based.

Despite the fact that you probably illegally jay-walked several times today (or did some other similar trivial illegal activity), the reality is that most people are law-abiding and will try to the best of their ability to follow rules as they are presented to them. (Teenagers versus parent's rules being the obvious exception to this, but that is a subject for a different book.) So, in the workplace, you might think that if there is an obvious rationale to temporarily ignore the risk management rules that employees will do so. However, the famous experiments of psychologist Stanley Milgram show that is not the case and why it is critical to think through the issues at hand before instituting a rules-based process.

In case you forgot your freshman psychology class, Stanley Milgram did a series of experiments to show how far people were willing to go against their better judgment and actually torture people in order to follow orders from an authority figure. In his famous experiments, Milgram had an authority figure who interacted with two research subjects, although one of the "research subjects" was actually an actor and a known part of the experiment. The subjects were told that the purpose of the study was to examine how receiving electrical shocks would affect learning ability. The "true" research subject was to ask the "actor" questions, and whenever they got an answer wrong, they would be given a shock. As the number of questions answered incorrectly increased, the shock would become larger and more intense. The "true" research subject was given a relatively mild shock at the inception of the experiment, so that they would have some appreciation for what the "actor" was feeling.

The "true" research subject was placed in a room, and in an adjoining room, the "actor" was supposedly hooked up to some electrodes. The research subjects could not see each other, but they could hear each other. As the experiment progressed, the "actor" would get more questions wrong and the "true" test subject would have to move a dial to increase the size of the shock they would receive. In reality, of course, the "actor" was not really hooked up to any electrodes, and was not receiving any shocks, but the "true" test subject did not realize this. As the shock increased, the "actor" would start crying out in pain, and eventually start begging the "true" research subject to stop. The authority figure, however, would remind the "true" test subject that they had agreed to participate in the experiment, that they were being paid to do so, and that it was important for the scientific research about learning.

The experiment, of course, was not to research learning abilities but was to see how far people would go in behaving a set of rules that were put in place by an authority figure. The results were disturbing at the time, and they remain disturbing to this day as the experiment has been replicated in many different forms. What Milgram found was that a large majority of the test subjects would continue to give larger and larger "shocks" to the "actor" even though they knew it was wrong to do so and even though they themselves were extremely stressed about doing so. In fact, many "true" test subjects continued to increase the size of the shock even after it appeared that the "actor" may have been killed by the shocks. When test subjects wanted to stop the experiment, the authority figure would remind them of the rules and that they had agreed to participate according to the rules.

The implications of this for risk management and process-based risk management are straightforward. If given a set of rules or processes to follow, and if the processes or rules are thought to be coming from an authority figure, then the reality is that people will be very reluctant to deviate from those processes and rules, no matter how much they believe that they should. Tomas Lopez was an outlier. As a risk manager, you cannot assume that people will use their judgment if a rule or a process is in place, despite how obvious it is that they should abandon the rule for that specific instance and use their judgment instead. If you are going to put in process-based risk management, you had better make darn sure that it is exactly what you want people to do, and that it covers all conceivable situations.

Simple, Complicated, or Complex (Again)

In Chapter 3, I mentioned the best-selling book by Atul Gawande titled *The Checklist Manifesto: How to Get Things Right* was discussed.[1] In dealing with situations that are simple, a checklist—or a process if you will—proves to be a very powerful tool or tactic for risk management purposes. A checklist, no matter how simple, or no matter how obvious the steps may be, helps to avoid mistakes and ensure that opportunities are captured. (It also deadens the brain for risk management, but this will be discussed more at length in Chapter 9: Can Your Risk System Be Too Good?)

Checklists for simple systems work because simple systems more or less follow rules and because they are robust. A slight to medium alteration to

---

[1] Atul Gawande, The Checklist Manifesto: How to Get Things Right, Picador, Reprint Edition, 2011.

the rules will still produce an acceptable outcome. Complicated systems, which adhere strictly to rules, are also amenable to risk management by processes, and in fact, complicated issues are the ideal type of situation that should be managed by a process or even by an expert system such as a computer.

More and more we see processes being used to risk manage complicated systems. In fact, we are going to extremes to manage these systems using processes. Because complicated systems are completely reproducible, do the exact same thing and get the exact same outcome, and as they strictly adhere to a set of rules or regulations, they can, in essence, be managed by processes that can be digitized. In other words, for complicated tasks, we can program a robot or a computer to manage them. After all, nothing follows rules more unquestionably than a computer. Problem solved! Computer-driven automobiles will save lives and likely get us to our destinations more quickly and in a more fuel-efficient manner. Automated security systems have replaced security guards at the entrances to the most secure parts of an organization. Computer systems might be hacked, but that is generally more difficult than bribing a guard or overcoming a guard by force. Robots are used in combat and in police work for the most dangerous of situations, and drones are used where it is too risky to fly manned surveillance aircraft. In many instances, the only use of humans in risk management is to give comfort to the general public who do not fully understand, or more likely, do not want to understand how much more efficiently and effectively that bots can manage most complicated tasks.

Complicated systems are so efficiently managed by bots that we create a new problem; that of trying to make every risk management issue a complicated one. We do this by assumption, or we do this by simplifying the situation. Sometimes, it is done by making heroic academic assumptions.

To take a basic example, we often assume that financial market returns are normally distributed. This allows the power of the well-developed mathematics of the normal distribution to be applied. Although we are still left with probabilistic outcomes, and not the completely reproducible outcomes of a true complicated system, it, at least, gives the appearance of such. Thus, we build pricing models and we value financial market securities with a false sense of great precision and accuracy. It then produces shock and surprise when a financial institution(s) suffers a large unexpected loss. A computer almost certainly produced a price, based on a complicated systems model, and no human checked to see whether the answer or the algorithm made intuitive sense as the mathematics were so elegant and academically verified. Regulators then set capital standards based on the models, and then later add a buffer factor to account for the unknowns. Processes- and rules-based risk managements do not do so well with unknown unknowns.

The issue with risk issues that are complex is that solutions are not possible. You cannot devise a process, an algorithm, or a set of rules to manage something that is complex and exhibits the property of emergence. As discussed in Chapter 3, the way to manage complexity is to think in terms of "manage, not solve," and also to think in terms of "try, learn, and adapt." There is no reproducibility in complex situations, and thus, there is no consistent way to manage or react to complex situations. Each instance is unique and inherently context-specific. Complex systems also tend to exhibit high levels of nonlinearity that produced very different outcomes based on very subtle differences. This is often referred to as the butterfly effect: the fact that a butterfly flaps its wings on the west coast can change the weather from sunny to a thunderstorm on the east coast a few days later.

Complexity requires judgment. You cannot digitize or codify responses to complexity. Complex situations require a manager who has the experience, the wisdom, and the courage to make judgments and the creativity and tenacity to continue to adjust their decisions. It requires real-time, context-specific management.

Risk situations that are complicated, and those that are complex, form the boundaries for the process versus judgment debate. To the extent that an organization has risk issues that are complicated, then they should utilize processes as their main risk tool. Conversely, those organizations that operate with mainly complex risk issues will require judgment as their main mode of risk management.

The practical reality is that for most of the time, risks will behave as if they are complicated. That is, they can be effectively managed by rules, regulations, or processes. The tricky thing is that occasionally, these seemingly complicated risks will take on patterns of complex risks. Then, a switch in management techniques needs to be made. Knowing when to make this switch from complicated thinking, and thus, process-based management, to complex thinking and judgment-based management is, perhaps, one of the most critical, yet, one of the rarest characteristics in effective risk management. It is doubtful if this trait can be taught, and it is doubtful if it is a characteristic that can be screened for in a hiring interview. It would seem that it is a trait that is possessed only by those who are highly intelligent and have the ability to live in the moment, although this assertion can be easily questioned and challenged. What is clear is that Tomas Lopez is one of the few who possess this ability to switch modes—or at least he did on the day that he performed his life-saving act of judgment.

The dichotomy between risks that are complicated and those that are complex illustrates the need to be able to understand and recognize the difference between complicated and complex situations. In terms of being able to effectively manage risk, it is unfortunate that so few important risks are complicated. On the other hand, in terms of providing rewarding careers for risk managers, it is fortunate that so many important risks are complex. Without complexity, risk managers could be (and should be) replaced by processes managed by robots and computers.

Playbooks Not Rulebooks

Jeff Swystun, President and Chief Marketing Officer of Swystun Communications, has a philosophy for creating great advertising that I believe also applies to risk management. That philosophy is "playbooks, not rulebooks." It is a great phrase to use to encapsulate the discussion about whether or not risk should be process or judgment-based. A playbook is a set of guidelines that should be roughly adhered to whenever practical. However, they are not rules that must be followed. When it is not practical to follow the guidelines, or when intelligent intuition says that a different path should be attempted, then the playbook takes a secondary role to responding based on the context of the moment.

I liken "playbooks not rulebooks," to coaching a team sport such as hockey. Before the game, and during a timeout, you will frequently see the coach with a whiteboard explaining how she or he wants the team to line-up and proceed with a tactical play or plan. Of course, the other team's coach is also doing the same thing and obviously both teams cannot, and will not, get their actual realized play to correspond to their respective plans. In fact, each team is doing their best to annul the other team's plan. Thus, each team goes back into action and implements the plan as best as they can, but it is likely very different from what the coach drew up. The team that is best at improvising and adjusting the plan according to some very basic principles is the team that generally succeeds.

Consider what would happen in the silly situation where a team is not able to follow the first step of their coach's plan, and thus, stops playing while the other team continues to play on while improvising. Obviously, the team that can only play to the exact specifications of their coach's plan is not going to do very well at all. In the early days of primitive computer chess programs, one could easily beat the computer by making highly unusual moves early in the chess game. For instance, moving the rook's pawn as one's first move would render many early computer chess programs useless as they could not compute how to react against such a strange move.

Playbooks, not rulebooks is, of course, a compromise position in the process versus judgment debate. Compromise solutions generally do not go over very well when asked to provide a solution or an answer. However, in our increasingly complex world, compromise solutions are often the best. Although board members, regulators, and stakeholders want assurances and documented proof of solutions, the reality is that business is not as black and white as we would like it to be. For some situations, a process works very well. For other situations, we have to fall back on judgment. At least, it makes risk management interesting.

Hiring Strategy

Before concluding this chapter, it is useful to take a few minutes to consider the hiring practices of an organization. No matter where an organization falls on the process versus judgment debate, its hiring practices should be consistent with it. If a company has a rigid process-based risk management system, then the people it hires should be amenable to such a system. One company I know of, who for obvious reasons will remain nameless, likes to hire people with low-level military experience, as they assume they are accustomed to following orders without question. If a company has a more judgment-based risk management process, then the people it hires should have the intelligence, maturity, and intuition to act on that judgment. The risk mindset of the hires needs to properly align with the risk system employed.

The second component to this is that the risk management training also has to be consistent with the style of risk management. If the risk system is process-based, then training needs to convey what the processes are, and perhaps, more importantly, why the processes are in place. When risk management is process-based, it is often incorrectly assumed that all one needs to know is the rules of the process. However, compliance will increase dramatically if the employees also know the why. Understanding the consequences—both good and bad—of compliance and of noncompliance helps the users who need to implement the processes appreciate the effects of their action. When someone understands why they are to follow a process, then they are much more likely to do so. They are also much more likely to appropriately deviate from the process when special circumstances dictate that they should do so.

Training for judgment is much more difficult than training someone how to operate a process. Often, it is easier to hire for judgment than to train for judgment. After all, one can only assess one's judgment with experience, and likewise, experience is the best trainer for judgment. As experience is the best trainer for judgment, judgment-based training needs to be much more experiential in nature than the typical three ring binder and PowerPoint deck training that is so common.

Concluding Thoughts

There is a place for process and there is a place for judgment in risk management. Rarely, however, does an organization require only one or the other. This creates difficulties and paradoxes for risk management. However, it also creates opportunities for those organizations that have the confidence to let their employees use judgment when appropriate. As a general rule of thumb, simple and complicated situations call for management by processes. Conversely, complex situations call for risk management by judgment, as processes are likely to backfire or at least produce unintended and unwanted consequences.

To get the best of both worlds, the philosophy of "playbooks, not rulebooks" is perhaps the best one to adapt. It is much harder to put into a risk manual, and it may be much harder to explain and justify to board members, regulators, and outside stakeholders, but it is often the compromise that is needed for effective risk management.

There is currently a definite emphasis on trying to make risk management as objective and as rules-based as possible. In part, this is driven by a need for regulators to be seen to be doing something. In another part, it is driven by society's false belief that in our knowledge-based economy that all problems can and need to be solved. Indeed, some risk issues can be solved, or at least effectively managed, through rules or processes, and in those situations, risk management should utilize process-based management as much as possible. However, there will always be a place for judgment as well. The fate of a drowning person may depend on the choice.

CHAPTER 8

How Do You Create a Great Risk Culture?

As someone who is trained as a scientist and then trained in finance, I am probably the last person who one would think of to write a chapter on organizational culture. My idea of the only time one should concern themselves with the human resources department is when they are handing you a specimen bottle for mandatory testing. I am not even sure I know how to define culture. One group of organizational experts seems to think it is what someone does when no one is watching. Another group of organizational experts thinks it is what one does when everyone is watching. I am not really quite sure which group of organizational experts is correct, but one thing I do know is that when I am conducting risk seminars, the issue of how to create a good risk culture is the one question that always seems to arise (after the question of "will I need a calculator for the seminar?").

Great risk management simply will not occur if the risk culture stinks. In the 1990s, Banker's Trust arguably had one of the best risk management systems going, but they failed in large part because their risk culture failed them. In the case of Banker's Trust, it was hubris and an arrogance around risk management that failed them. Their risk management processes were world-class, but the culture was lacking. In most firms, the risk culture stinks for very different reasons. People are afraid of risk. People are skeptical of risk. People believe the risk department has the sole function of keeping their foot on the brake in order to kill any progressive idea or project. People want to avoid the risk function, as it is seen as a bureaucratic jumble of red tape and meaningless paperwork. In other words, risk stinks because the culture around risk stinks.

Perhaps, I am more than a little unstable mentally, but I think risk is cool. I think the risk function has the tools, techniques, and tactics to allow an organization to extend itself and really stretch for new goals and heights. I think that risk management is one of those true areas of management, where both the art and the science of business come together and allow the really talented people to strut their stuff. As such, risk culture should be cool, hip, and progressive, instead of the traditional corporate career graveyard that traditionally it has been. It should be the area of the organization where the people with the most ideas, the most creativity, and the most energy congregate and want to work in. Let's get started exploring some ideas on how to create not just a good risk culture, but a great risk culture.

Culture Is the Middle

A lot of the reading about risk, and risk culture that I have done, and most of the people I have discussed the topic of risk culture with, say that risk culture starts at the top. I respectfully disagree. The rest of the literature and the rest of the people seem to think that risk culture starts at the bottom of the organization. Again, I respectfully disagree. I think that culture, and in particular, risk culture is made or broken in the middle.

In some sense, I realize that culture does absolutely begin at the top, but it is only at the lower levels of the organization—where the rubber meets the road—that culture ultimately gets implemented. A culture, however, needs to be consistent, and if the top and the bottom of the organization do not "get" the same "memo," there is a disconnect. That disconnect is the middle manager.

The middle manager in any organization is in a very unique position. Those at the lowest rungs of the ladder realize that they have a lot of room to move up before things start to get serious. They are also probably content to just have a decent job at their given stage of their career. Those at the top of an organization have, in some way, "made it," and thus, enjoy a freedom and flexibility that most people in the organization do not get to experience. Those at the top are also generally quite out of tune with what is happening at the front lines. (Of course, this is the premise of the popular television show "Undercover Boss" that puts senior executives in frontline positions and highlights their ineptitude at doing the very tasks that their companies are in business to do.) The middle manager is squeezed between these two groups. The middle manager likely has aspirations to move up, and thus, is hesitant to do anything that will jeopardize their placement in the upper echelon. They are also the ones who are on the hot seat when the rubber does not meet the road properly, but as they are one step removed from the front lines, they have limited ability to affect their own fate. The middle manager has a quandary and a unique risk management problem as well. Do they play it safe and avoid a misstep that could get them fired, or do they take chances and try things in order to get ahead? Their career choices are a risk management exercise, in a nutshell.

Almost by definition, middle managers are the epitome of the "organization man or women." If they had an entrepreneurial mindset, they likely would have left the organization once they reached middle management. A few of them may be stars and shooting for the top, but in my experience, many of them quickly realize the odds are stacked against them from reaching the upper reaches of management, and thus, they settle for a career as a middle manager. This creates a mindset that is focused on not rocking the status quo. It also creates a mindset that any risk—upside risk or downside risk—is bad and to be avoided at all costs. There are obviously exceptions, and hopefully, in your organization, the exceptions are many, but middle management is not the place to find a healthy and positive attitude toward risk.

As middle management is where the connection is made from the upper reaches of management to the frontline workers in the trenches, it is also where the culture of risk gets set. If you have middle managers with an unhealthy and negative view of risk, then the organization will have an unhealthy and negative culture around risk, despite the proclamations of the board, the CEO, or the entire senior staff. It is a tough problem to crack.

There are a few central causes to the middle management problem. The first is the process for promotion. In any organization, you generally get hired for some type of technical skill, and perhaps, that skill is management itself, but more likely, it is some type of skill that is best designed for complicated tasks, such as accounting or engineering. Someone who demonstrates proficiency in accounting or engineering, for example, then gets promoted to middle management. However, this is without any real management experience, or even evidence of managerial skill. This, of course, is a well-known problem as accurately and humorously outlined as the Peter Principle.[1] However, not only does promotion occur without the employee providing evidence of managerial skill, but it also occurs without the employee providing risk management skills—except for risk management skills in their technical expertise.

The second cause of the middle management issue with risk is the risk training that they receive. Almost all of the risk training is in terms of regulatory aspects of risk, and the risk processes—one might say risk bureaucracy—that is in place at the organization. This training generally occurs at the inception of the manager's promotion when the attitude of the new middle manager is most impressionable. The impression made is that risk is a process-based, brain dead jumble of bureaucratic red tape. No wonder most middle managers develop, and thus, implicitly promote, a robust dislike of the risk management function.

Ironically, the senior managers who I have had the opportunity to work with tend to have much healthier attitudes toward risk. Admittedly, many do not agree with my definition of risk, and many believe that risk management is only to protect against bad things happening. The big difference is that at least they consider the question of what risk is, and what the role of risk management is. Additionally, while their technical understanding of risk may be weak, and in the current regulatory climate of the Sarbanes-Oxley Act, their lack of a strong technical understanding of risk is a source of fear for them, these senior managers do tend to have a very good intuitive understanding for the two-sided nature of risk and the inherent uncertainties of risk. Perhaps, this intuitive understanding of risk is what makes them qualified for senior management as opposed to those lower-ranked employees who have a better technical understanding of risk.

The culture of any organization obviously begins at the top and conveying that culture from the top throughout the organization is always

---

[1] Laurence Peter and Raymond Hull, "The Peter Principle: Why Things Always Go Wrong," William Morrow and Company, 1969. To remind the reader, the Perter Principle states that a manager will get promoted until they reach their level of incompetence. This implies that all senior managers are, thus, operating at their level of incompetence.

going to be a challenge—no matter what form of culture we are talking about. However, when it comes specifically to risk culture, I believe that there is a special problem in getting the message from the top of the house to the front lines because of the unique career-based risk issues facing the middle managers. When it comes to risk culture, middle managers control the knobs on where judgment versus processes are set, they control the knobs of what the definition of risk is going to be, and while they likely do not control the risk metrics used, they control how those risk metrics are interpreted and utilized. In essence, middle management controls the risk culture.

Banish Risk as the "Department of No"

Perhaps, there is no more fruitful step toward promoting a positive risk culture than banishing the thought that risk is the "Department of No." The risk management function as the "Department of No" is an obvious downer. This has been a major theme running throughout this book, so I will not beat the dead horse once again, but leave this section intentionally short to drive home what should be an obvious, but almost always overlooked point in creating a positive risk management culture; banish risk as the "Department of No!"

People Not Processes

Prioritizing people, not processes should be an incredibly obvious step toward creating a great culture, but apparently, it is not. Culture is about people. Thus, if you want to have a great culture, you need the risk management function to have the focus on people not processes. This includes thinking about the people first in the design of risk management practices, it includes the training around risk management, it includes the rewards and motivations around risk, and includes people in the decision making around risk. It also includes hiring for a certain risk management attitude—attitude that you want your risk culture to be.

When risk management becomes an engineering problem, or a regulatory exercise, then the focus on the people often gets lost in the minutia of the processes. The issue is that, as discussed in Chapter 4, risk is rarely caused by complicated engineering issues, but much more frequently has complex people-related causes. Furthermore, it is almost always people that have to implement risk management. Thus, it just makes sense to design risk management with people first. However, it is almost done the other way around. This makes one feel insignificant, unimportant, and thus, unmotivated. Basically, when you realize that you are playing second fiddle to a process, it is a soul-destroying realization.

Take a very common risk management process: that of the safety announcements on airplanes. How many times have you sat through a demonstration of how to put on your seat belt in the airplane? Did you feel like you were back in preschool day care and the teacher was telling you it was nap time? Did you pay attention to the demonstration with rapt attention or did you try to show that you were purposefully ignoring it while all the while trying to block out the annoying litany—which you have probably memorized after your first three flights? Did you mumble under your breath "this is stupid"? Do you feel you are in a positive culture of safety or a pawn of a cruel hoax imposed by some brain dead bureaucrat of one of the myriad of transportation agencies? Do you believe that the safety message is to help you improve your flying safety or was designed as a legal protection strategy for the airline? Do you get cynical about the whole thing when the airline simply puts on a recorded message while the flight crew stands at their regulated spots going through a lackadaisical mime routine, showing you how to buckle and unbuckle your seatbelt and tighten the oxygen mask "around your nose and mouth"?

There is one airline that I fly frequently who take a different approach. Knowing that their customers are, for the most part, frequent flyers, they make an ongoing joke and allow their flight crew to inject their personalities into the safety announcement. I am not going to name the airline for fear that some bureaucrat sees this and hunts them down and forces them to stick to the script. The point is the flight crew is making an effort to show they realize that there are people, and not bots seating in the seats. The crew are also making an effort to show that they believe their customers are not idiots who never learned how to fasten and unfasten a seat belt.[2] It is a small thing, but it gives you a different attitude toward the demonstration. A vastly greater proportion of flyers listen to the announcement, and when they do buckle up, they do so with a positive attitude, rather than "here we go again" attitude that they do with the more usual litany. A realization that people are in the seats and also that it is people (flight crew) who are giving the message allows the safety message to be about people and not just the regulatory aspects. It leads to more interest in the message from both the crew who have to improvise—a bit—and to the flyers who listen with interest to see how interesting, and perhaps, entertaining the crew can make the mundane instructions. More interest leads to more compliance, more awareness, and more acceptance. Perhaps a trivial example, but one that illustrates the power of putting people before processes.

It is very basic, but risk management processes are in place for the good of people—employees, customers, suppliers, the general public, and so on. People, stakeholders of an organization, do not exist for the risk management processes. However, it often feels like it.

Risk Happens

Part of a risk management culture is very simply admitting that risk happens—both good and bad. Risk happens despite everyone's best intentions and despite the strength and resiliency of the risk management system. To create a great risk culture, a firm needs to admit that risk happens. It needs to celebrate the good risks, and try to the best of its ability to learn and to improve from the bad risks. It needs to stop punishing the occurrence of risk—unless it was a deliberate and willful act to trigger a negative risk outcome.

Too often, the culture is set up to let those know that if they are the cause of a risk outcome, then there will be consequences, and the

---

[2] I actually was on a flight once in the Middle East where the person seated next to me was flying for the first time. They were the only person I have encountered in my entire life who did not know what the seatbelt was or how to fasten or unfasten it. Some bureaucrat will claim this is the exception that shows why it is still necessary, but I suggest they need to get real.

consequences will not be nice. This leads to a culture of fear and loath-ing toward risk management. Perhaps, more seriously, it leads to hiding events from risk management.

I can clearly recall an incidence when I was a young lad—about five or six. Before you think my parents were willfully neglectful, I should point out that it was a different era, and as kids, we were actually allowed to do such risky things such as play outdoors, and—horrors of horrors—our parents would occasionally even leave us alone at home during the daytime. It was during one such day that my parents left me home alone and I knew there were cookies in the uppermost reaches of the kitchen cupboard. I also knew that if I got a chair that I would be able to climb up on the cupboard and pilfer a few cookies. Needless to say, the temptation was too great. What I did not quite allow for in my plan was the newness of the expensive countertop that I was now climbing on, and in my climbing, I chipped the corner. Knowing the risk attitude of my parents, I knew the outcome was not likely to be good. Thus, I got some glue and tried to glue the chipped pieces of the countertop back in place. I thought I did a great job, but when my parents came home, my mother caught her new dress on my handiwork and ripped the new dress that she was so proud of. Not only had I been found out, but my relatively small problem had turned into a much bigger problem. Needless to say, I was grounded and my lust for cookies has been diminished for life.

The reason behind the story is straightforward. If mistakes are not tolerated, then mistakes will be hidden. Furthermore, when mistakes are hidden, they almost always grow into bigger mistakes. Almost no one sets out to make a mistake or cause a negative risk event. However, mistakes and bad risks occur. If mistakes and risks are not tolerated, then they will be subject to attempts to hide them or cover them up. It is like someone being in denial about having cancer. Just like cancer, the cover up will almost always lead to a series of further mistakes and mishaps and the problem will continue to grow and fester.

Ironic as it sounds, mistakes and risk events—both good and bad—should be celebrated as learning events. In essence, risk events are teach-able moments. The risk culture has a choice about what to teach. It can teach that risk is evil and risk must be avoided at all times—both good and bad risk—or it can teach tolerance and learning and improvement. Risk events should be teaching moments, not scolding and blaming moments.

In terms of athletic ability, I am (and was) quite mediocre, but somehow I managed to be quite successful in organized sports. How, you might ask, given my limited athletic ability. The answer was simple—I tried hard to never make the same mistake twice. I tried to learn from my mistakes on the playing field, rather than berate myself for my mistakes. It was quite likely I was going to continue to make mistakes and to lose points. However, by concentrating efforts on not making the same mistake twice, I not only avoided repeating mistakes, but I was forced to think of different ways of doing things. Thus, mistakes became opportunities to improve myself during a match, rather than a method for getting my spirits down. The same philosophy can also work for a risk culture. If the emphasis is on not making the same mistake twice, and if there is a tolerance for making a new mistake, then the organization not only accelerates its learning, but it also improves the risk culture.

Having a risk culture that recognizes that risk happens allows everyone to take a collective breath. It lowers the fear of making a mistake, which, in turn, produces fewer mistakes. When the atmosphere is so tense, and there is a paranoia about doing something wrong, the ability of people to think and to respond properly is diminished. Worse yet, people may get paralyzed for doing the wrong thing, and thus, do not trust their instincts and wind up doing nothing, even when action is called for. It is far less stressful, and far more productive to take a collective organizational breath and freely admit and realize that "risk happens."

Risk Training and Culture

What did you learn from your organizational training on risk—that is, assuming that your organization has risk training? Was the risk training fun, or was it a boring litany of thou shall not? Did you leave with positive or negative thoughts about risk? Was the training based on knowing things about risk or in how to think about things surrounding risk? Do you even remember anything from your training?

If a firm is to have a positive culture about risk, the training about risk needs to take on a positive tone. The training should emphasize the risk philosophy of the firm, the fact that risk management is an active part of the competitive advantage of the firm; that risk management is not the "Department of No"; and that risk management is inherent in everyone's role.

A second important aspect of the training is that it should focus on understanding and appreciating risk, not just on knowing the textbook aspects of risk. Knowing and understanding are two different things. If the training is focused on knowing about risk, then risk is seen as a complicated system. If the training is focused on understanding and appreciating risk, then the complexity of risk and the creativity in managing it comes to the fore.

My colleagues and I once conducted a daylong risk training simulation exercise for a large international organization. Over 100 managers of this organization had been through a series of risk training workshops over the previous year led by well-known academics and consultants. The duration of the training spanned a full 10 days. The purpose was to make the organization state-of-art when it came to risk management. The daylong event was to end with each of the groups making a series of presentations for their analysis and solution of the issue that we designed for the simulation. The senior managers were to be there for the final presentations. The whole meeting was designed to be the capstone event for this major training initiative.

The simulation we designed was closely based on an actual situation that had occurred roughly 15 years previously to a similar company. For the simulation, we basically updated the scenario and used newer data and a slightly disguised product for the participants to conduct their analysis on. We were slightly concerned that someone would recognize the true story we had based the simulation on, but no one did. In short, the simulation was based on designing the risk management of a new product launch. What happened in the real-life situation was the company did everything correctly in terms of the complicated processes, but they neglected how to deal with the media and the reputational risk and the social risk management of the project. In the real-life situation, the risk management broke down and fell apart with a call from the media.

In the simulation, which ran for six hours, my colleagues and I worked with each of the groups. I was very impressed with their knowledge of risk, and each of the teams was doing a great job of preparing a presentation that showed that they understood risk and had mastered the training that they had been given. The problem was they demonstrated a lot of knowing and not a lot of thinking. When everyone convened in the main hall for their presentations to senior management, they had on each of their desks the same memo from the media that had occurred in the real-life situation, and just like the real-life situation just over a decade ago, the risk management plans fell into disarray, and furthermore, they fell into disarray in a hurry.

What this example illustrates is that training based on knowing things about risk is ineffective. Training about risk needs to be on thinking about risk. Knowing "stuff" and being trained to think are two very different things. While it is admittedly good to know things, a simple knowing does not imply understanding or knowing how to think about a situation. In part, knowing stuff implies that risk is a complicated system. So, therefore, if the majority of your risks are complicated, then by all means focus on training to know. However, if like most firms, your organization's risk arises because of complex issues, then you need to train based on thinking and understanding and creativity and appreciation for the complexity—a very different approach.

Training for knowing things implies an arrogance that risk is something to be optimized. However, if risk can be so easily optimized, then the future of risk managers is bleak because, as previously discussed, we will all be replaced by computers who are much better at optimizing complicated things than we as humans are.

Risk Should Be Fun and Creative

Where do the creatives in your organization exist or migrate to? (If you do not have creatives in your organization, then you really should start looking for a new job.) Do the creatives, those with an entrepreneurial mindset, those who like to connect the dots, and those who like to search for new dots, do they exist in your risk function?

Admittedly, there is a lot of engineering and math and best practices in risk management. In other words, risk management has a chuck of science associated with it. However, I would argue that the really important and valuable risk management activities are those that are much more art and much more creative and much less science than we normally ascribe to risk management. It is hard to be creative when the mood is doom and gloom. Much more creative productivity occurs in an atmosphere of fun and optimism.

Risk should be fun, and risk should be creative. Earlier in the book, I stated my first law of risk management: the mere fact that you acknowledge that a risk may exist automatically increases the probability of it occurring and its magnitude if it is a good risk, while also automatically decreasing the probability and severity of it occurring if it is a bad risk. Thus, to be good at risk management, one needs to come up with lots of creative scenarios. With risk management being buried under processes and mounds of data collection, and heaps of regulation, it is hard for even the most creative of people to be inspired.

As a university professor, I have the vicarious pleasure of seeing my best students set off on their career's at the end of each school year. I am always surprised by how companies recruit who they believe are the best students. Most of the time, companies recruit the worst students. I wish companies could see the process from my point of view and hear the conversations I do coming from the truly great students with the most potential.

As someone who teaches risk management, I am often asked by recruiters and human resource professionals who the best students are and who they should look out for. When I ask the criteria they are looking for, they almost never ask for creative, or fun, or forward-looking, or ability to deal with ambiguity. Instead, they claim they want those who know the math of risk the best, and are organized. (Really—organized is the key skill you are looking for in a junior risk manager?) What I observe by following the success of my student's careers though is that it is those who are creative, who see the possibilities and the potential of risk management, and who enjoy thinking about risk management problems and opportunities are those that do the best. Those who are focused on the techniques of risk management might have the easiest path getting hired, but they generally stall the earliest in their careers or burn out the quickest.

In short, many organizations that are recruiting for their risk management department make it seem like the most brain dead and boring of departments. In fact, they should use a form of reverse hiring—anyone who expresses interest in applying for a job after listening to the company's pitch at their information sessions should be eliminated from contention.

One company—a respected global leader in risk management consulting—got it right in my opinion. On campus, instead of having the traditional corporate information session, they held a board game tournament. The game they used was of course *Risk*. The interesting thing was that the engineering types thought the idea was stupid. The students who were creative and looking to avoid a boring job at a boring company thought it was a great idea and eagerly participated. I guess it comes down to what type of people you want to try to recruit, but then again it comes down to what culture you want to create around risk management.

Concluding Thoughts

To reiterate a point I made at the beginning of this chapter, I am probably not the one most people would choose to pontificate on creating a great risk culture. However, having worked in several different types of risk cultures, and consulting for many more, I am a believer that the risk culture of an organization is absolutely key to having great risk management.

Historically, it was often the case that risk management was the department in which an employee who had reached their usefulness as a manager, but was not yet ready to retire would be placed. Risk was thought of as a sleepy place where you needed experience about how things worked—and how they did not work. The work was thought to be dull, relatively unimportant, and not very dynamic. In essence, the risk department was akin to a corporate retirement home. I would argue that this concept is about as quaint as a rotary dial corded phone. It is from another era that is not returning.

In contrast, I believe that risk is about the most dynamic, important, challenging, and valuable function in an organization. Making that message clear, and constantly reinforcing the message that not only bad risk needs to be managed, but good risk as well, is critical for a firm's success.

CHAPTER 9

Is Your Risk Management Too Good?

The current thinking among many pediatricians and child-rearing experts is that we have developed a generation of overmanaged "bubble-wrapped" kids. While keeping our offspring relatively free of broken bones, knee scrapes, and misery, inducing infectious diseases, is quite admirable; it may be doing more harm than good in the long run. In a 2015 position paper,[1] a diverse group of child health and safety experts issued a position paper that stated, "Access to active play in nature and outdoors—with its risks—is essential for healthy child development." The paper continued to warn against "'Hyper-parenting,' 'invasive parenting,' or 'intensive parenting,' in which a climate of 'inflated risk' leads parents to micromanage all aspects of their children's lives in an effort to protect the child from adverse experiences." In essence, it was a call for parents to stop bubble-wrapping their kids and let them play freely with all of the resultant risks that this might entail.

Is the same happening in corporate management? In other words, are managers becoming so risk-adverse and so conscientious in designing and implementing risk management strategies that they are in effect creating "bubble-wrapped" corporations? Furthermore, is this risk-centric management paradigm setting up companies to be unable to deal with the inevitable risks that are inherent to being in business? Has risk management in effect become "too good"? The phenomena known as risk homeostasis says that the answer is yes!

---

[1] Brussoni, M. et al., "Position Statement on Active Outdoor Play," International Journal of Environmental Research and Public Health, 2015, 12, 6475–6505, available at http://www.mdpi.com/1660-4601/12/6/6475.

Why Is Risk Management So Strong?

There are four reasons why risk management is so prevalent in the current mindset of corporations and the corporate culture in general. The four reasons, which are interrelated, are: (1) a natural dislike of uncertainty, (2) the culture created by the marketing and politics of fear, (3) the legal and regulatory framework of risk, and (4) managers have an almost innate need to believe that they are doing something useful in part to avoid the onset of the imposter syndrome.

It is quite natural to dislike uncertainty, although admittedly not all do. Some people like uncertainty—they like not knowing what is coming next, to essentially live their life going from surprise to surprise. However, these people are in the minority, and most certainly, the stock market does not like surprises. Even good surprises such as an unexpected jump in earnings can be met with less than an enthusiastic response.

The main reason people dislike uncertainty is that it lays waste to the best designed plans. The old management axiom of if you want to see God laugh all you have to do is tell her or him your plans holds whenever risk and uncertainty are present. The degree of success of strategic planning, and even tactical planning, is generally inversely related to the amount of uncertainty. The greater the level of uncertainty, the less effective plans tend to be. A high level of uncertainty implies that managers need to think and react in real time, which, in turn, can be a creator of stress and of course mistakes. Uncertainty, of course, also implies that things may turn out better than expected, but the natural pessimism that many managers seem to possess means that upside uncertainty or risk is often discounted. The assumption seems to be that Murphy's Law—that anything that can go wrong will go wrong—is the dominant mindset.

While it is natural to want to be able to plan, to foresee the future and have a good idea of what is going to happen, it is not reality. The world of business is inherently risky and uncertain. That is also a very good thing, for without uncertainty, there would not be a need for managers and decision makers. Nor would there be a need for experts. Without uncertainty, it would be a relative straightforward task to calculate the optimal path and execute a plan rather simplistically. For this reason, it is fortunate that the world of business is not as certain as the natural world of physics.

Uncertainty causes stress, frustration, extra work, and it demands focus, flexibility, and determination to deal with ambiguity. Given this, and the constant demand from investors for predictable returns, it is easy to see why risk management has become a central activity of manager's, directors and regulators.

Building on this natural tendency to want to avoid uncertainty is the marketing of risk. In his book, *Risk: The Science and Politics of Fear*,[2] author Dan Gardner highlights numerous cases of where a fear of risk has been exploited for political and marketing means. Gardner highlights how fear of uncertainty has been used to "market" climate change, political campaigns, security systems, antibacterial hand lotion, and a host of other consumer products or political concepts. Fear of change, and more particularly, fear of a loss is a very powerful motivator, and thus, provides an extremely effective basis for a marketing message.

All of this, of course, has played into the hands of the burgeoning legal and regulatory framework that managers now find themselves operating within. Sarbanes Oxley (SOX) regulations, Basle III regulations along with Dodd Frank rules for the financial industry, as well as the omnipresent tort legal framework are an increasingly large set of rules and threats that managers need to be cognizant of on a daily basis. While regulations have always been a necessary part of almost any well-functioning economy, their presence seems to be increasing at an unnecessary exponential pace.

While there are many other reasons, some good, some not so good for having a strong culture of risk management, the final major factor is that professional managers have an innate need to believe that they are doing something to justify their position as manager. Just as a parent naturally wants to do everything in their power to protect their child from unnecessary harm in order to consider themselves a good parent, a manager likewise needs to conceptually believe that they are doing everything in their power to protect the assets entrusted to them by the firm's stakeholders. While the glory of implementing a profitable strategy can accelerate a management career, the reality is often that the glow from such success

---

[2] Gardner, Dan, Risk: The Science and Politics of Fear, Virgin Books, London, 2008.

fades quickly, while the disgrace of a major miscalculation stays with one's reputation forever.

The economist John Maynard Keyes supposedly once stated that "worldly wisdom teaches that it is better for reputation to fail conventionally than to succeed unconventionally." The current convention is to have an unassailable set of risk management protocols, no matter what the cost, or what the effect is on the efficiency or growth of the firm, and thus, risk management rules.

The well-documented imposter syndrome plays into this. A manager who does not take every precaution is likely to be accused of being negligent in a career harming way. To justify their corporate existence and paycheck, managers often believe that they are first and foremost to protect against losses and in turn this gets translated into making sure that the firm has a strong risk management culture and system.

Harmful Effects of a Too Good Risk Management System

So, how can something as intuitively and naturally good as wanting to avoid risk become a bad thing? Is it really possible to have too much of a good thing? The answer is an unqualified yes! There are a large number of ways that it is possible to have too much risk management or too strong of a focus on risk.

The first and most counterintuitive effect is what is known as risk homeostasis. Risk homeostasis is a well-documented phenomenon that states that people will react in such a way that they take riskier actions to counteract the risk prevention mechanisms in place until the overall level of risk is the same or even greater.

For example, assume that you are about to go driving in a strong snow storm and the roads have not yet been plowed and you know they are very icy and slippery. Assume that you need to drive to a location in a small compact budget car that does not have snow tires, nor does it have any snow-based traction control systems. Now suppose that you make the return trip under the same wintery conditions, only now you are driving a large luxury sports utility vehicle with four-wheel drive, that has snow tires as well as advanced traction control systems specifically designed for driving in snowy and icy conditions. Does your driving style change? The statistics say yes, as paradoxically you are more likely to be in an accident driving the luxury sports utility vehicle that has all of the advanced safety systems. With the sport utility vehicle, you will drive more confidently, and thus, less cautiously. This change in driving behavior actually increases your probability of being in an accident, despite the added safety systems. In fact, if you have such a vehicle, you are more likely to make a dangerous trip than you would if you had an econobox car that you knew would be tricky to drive in the hazardous winter conditions. For this simple reason by itself, you are actually increasing rather than decreasing your risk.

As another example, consider crossing a busy street in a large city. You have a choice of crossing at a designated crosswalk with a crossing signal, or you can dodge traffic and jaywalk in the middle of two intersections. When you jaywalk, you are more likely to be careful to look both ways, actually look out for cars and be more careful in your steps in crossing. You also will not be looking at your smartphone while jaywalking. However, when you cross at the crosswalk, you are much more likely to cross without checking to see whether all traffic has stopped, and you are also more likely to cross while checking your handheld device. If in the crosswalk, you are also likely to be oblivious to a vehicle that failed to stop for whatever reason for the traffic signal.

Risk homeostasis is not limited to specific safety mechanisms, such as traction control or crosswalks. It can also extend to a risk culture. Extending the example of crossing a busy city street, take for instance, the case of crossing the street in my home city of Halifax Nova Scotia. In this city of approximately 350,000 on the east coast of Canada, there is a strong culture of cars stopping for pedestrians who are waiting at a crosswalk to cross the street. Visitors to Halifax frequently comment on it, and in fact, it is a local joke that cars will stop even if it appears that a pedestrian might be even "thinking" of crossing the street. Despite this respectful and safety conscious culture, there has been a rash of car–pedestrian accidents in crosswalks, so much so that the government has increased the fines for pedestrians crossing inappropriately to $700—the same fine that exists for cars that fail to stop for pedestrians. The problem is a case of a risk culture causing risk homeostasis. Pedestrians have become so accustomed to cars stopping for them that they no longer bother to "look both ways" before crossing at an intersection.

Contrast the situation in Halifax with that of Drachten in the Netherlands. According to Taleb,[3] the town of Drachten had a similar problem with car–pedestrian accidents, but their solution was unique and counterintuitive, but it worked. The town of Drachten simply did away with virtually all crosswalks, and the number of automobile accidents dramatically reduced. Cars became more aware of pedestrians attempting to cross at random places, and pedestrians, in turn, become more cautious when crossing without the benefit of a designated crosswalk. Without a risk system (crosswalks) and without standard rules for street crossing, both cars and pedestrians become more attuned to the risk, and thus, the problem solved itself organically.

With a strong risk system, or a strong risk culture, the individual accountability for risk management is greatly diminished. The onus lands on the systems, or the rules and regulations, to manage the risk, rather than the responsibility of the individual. Without the individual account-ability, risk homeostasis sets in and risk often overall increases.

In a corporate setting, employees can become complacent or worse yet "lazy" in their risk thinking and their risk awareness. In essence, man-agers and employees can become disengaged from risk management, just as the pedestrians of Halifax through the car–pedestrian culture became disengaged from the dangers of crossing the street. This is particularly true if employees believe that the risk management function is so strong that it is not worth trying to change it, or work against it for the better-ment of the company.

Another side effect of being risk-lazy is that employees will not be looking out for good risk, that is unexpected good outcomes of a risky situation. In many organizations, the risk culture stifles creativity and gen-erates a momentum of its own for the status quo. In essence, employees quickly learn what the risk system allows, and just like Pavlov's dogs, they react in a way that confirms to the risk system, rather than thinking independently.

---

[3] Taleb, N., Antifragile: Things That Gain From Disorder, Random House, New York, 2012.

Of course, in certain cases, they may take the exact opposite view and work to sabotage or work around the risk system. If the risk system is assumed to be strong and reliable, then a rogue employee with nefarious intents can use the firm's confidence in risk management to their own purposes. For instance, there are several well-publicized cases of financial traders who learned to "cheat" their firm's extensive risk systems for their own benefit, but also to the ultimate detriment of the firm. Ironically, believing that the risk system is infallible makes it that much easier to cheat it.

Additionally, if a risk system or culture is too strong, employees and managers will not be consciously or unconsciously thinking about how risk management could be better or how risk technology could make things more efficient and effective from a risk standpoint. This is a very important point that will be discussed in more detail later in this chapter.

An obvious and explicit downside to having too strong of a risk system is the costs, both those that are explicit as well as the often uncalculated implicit costs and opportunity costs. Having a strong risk management system has explicit costs, such as the systems, and the personal required to implement and maintain it. As one risk manager told me about their experience implementing Sarbanes Oxley (the government regulations imposed on corporations to prevent corruption and financial fraud), that in estimating the costs, one should take their worst estimate in terms of explicit costs, as well as their worst estimate of time required and then quadruple them in order to get the requirements to being 50 percent implemented. Regulations such as SOX and Basle III regulations for financial institutions are the main drivers behind the vast number of employees being trained and employed as risk managers and the concurrent rise of Masters in Risk Management programs that are proliferating. It is an interesting thought experiment to consider the valued added to an organization if this cost and the talent associated with regulation were simply employed in risk management in a manner that the organization felt was most useful, rather than in the manner that regulators require.

However, rarely does an organization calculate these direct costs, and even more rarely do they calculate the return on the risk management investment. In a widely cited case study, John Fraser, the former Chief Risk Officer at Hydro One (the major electrical distributor for the

Province of Ontario), gives the example where capital allocation at Hydro One was based on "risk bang for the buck."[4] Risk bank for the buck is a process whereby divisional managers at Hydro One have to quantify the amount of risk their unit had and the return per unit of risk reduction. Capital allocation is based on where the capital would have the greatest return per unit of risk reduction.

Regulators suffer from three drawbacks as adjudicators of risk. The first is that they need an objective set of benchmarks for which they can fairly and consistently apply their standards across the organizations that they are responsible for. The second is that they do not have "skin in the game" as coined by author Nassim Taleb.[5] The third is that they are often responding more to political pressure than real-time concerns. This, of course, is most evident after a crisis occurs, and thus, we get the often voiced criticism that regulators are always managing the last crisis, not the more critical forthcoming crisis.

Finally, a too strong risk management system or culture creates an atmosphere of paranoia, where employees are afraid to attempt the slightest of innovations and managers manage to the rule rather than to the situation. This creates a stifled workplace and leads to delays in product innovation and general overall efficiency and effectiveness of the work environment. It most definitely is possible to have too strong of a risk management system.

Steps to a Healthier Attitude Toward Risk Management

The purpose of the preceding arguments is not to say that risk management is bad or unnecessary, but rather to illustrate that the tail may be wagging the dog. It has also been to show that excessive risk management is not benign, but rather that it may be having a harmful effect on the efficiency and efficiency of the organization in a plethora of ways, both

---

[4] Aabo, T., Fraser, J., Simkins, B., "The Rise and Transformation of the Chief Risk Officer: A Success Story on Enterprise Risk Management," Journal of Applied Corporate Finance, Winter 2005.
[5] Taleb, N., Antifragile: Things That Gain From Disorder, Random House, New York, 2012.

explicit and implicit. You really can have too much of a good thing in risk management.

The first step toward more effective risk management is to clearly recognize and communicate what the goal of risk management within the organization is. As discussed previously, frequently risk management is seen as the "Department of No," when instead it should be seen as the "Department of How to Do Things Better." This, of course, immediately raises the question what exactly it means to say "better."

By rethinking and reframing the role of risk management, an organization can dramatically increase the effectiveness and the value of its risk function. The one difficulty of doing so is that the change in tone by the organization will not by itself change the demands and expectations of other stakeholders, and most significantly, the demands and expectations of regulators. The regulators have an interest in only downside risk, that is, they will only receive publicity when adverse events happen, and not when something goes right in an industry. Thus, convincing regulators that risk has a positive side in addition to a negative side is a very difficult sell. (Ironically, politicians should have an interest in upside risk as if a company or an industry is doing better than expected then the economy will be doing well. However, as discussed previously, fear sells, and thus, negative politics keeps the focus on the negative aspects of risk.)

The asymmetry between the objectives of regulators who want the focus to be on negative risk and that of an organization that wants to manage both the positive as well as the negative types of risk can be reconciled in a fashion akin to accounting. Most for-profit companies in effect have three different sets of accounting statements. They have a set that they use for financial reporting that are subject to accounting standards board's rules. They also have a similar set of financial statements that are used for tax reporting purposes. However, few organizations use these mandated and regulated statements for actually managing the business. Instead, they maintain a third set of statements for managerial purposes, and thus, the term managerial accounting. In a similar fashion, companies could maintain a set of risk records for the regulators, but also maintain a set of risk measures and metrics for managing both the upside and the downside risks that they face.

A second paradigm shift needed in risk management is the returning to the reality that all business decisions are inherently risky. It gets back to the central tenet that states "no risk, no return." This statement seems so obvious as to not merit mention, but in the current culture of excessive risk management, it is too often forgotten or lost in the process. With excessive risk management, the mantra seems to be that all risk is bad. This is pervasive not only from the legal, regulatory, and political realms, but has also entrenched itself within the corporate world itself. It is, perhaps, not a coincidence that entrepreneurship, which is most often based on new paradigms that have yet to attract the attentions of authorities and regulators, is seen as the corner of business in which the risk-takers dwell, while the established corporations are seen as the conservative realm for those more interested in professional management. Risk is inherent in business, and furthermore, it should be inherent in business.

Creating the proper culture around risk is imperative. Risk, and in particular, good risk, or upside risk, needs to be embraced. Even downside risk needs to be embraced as an opportunity or a challenge to create and engineer processes to make the risk of the situation more acceptable, and in doing so, create a competitive advantage.

One method for creating a proper risk culture is to foster an environment where honest and well-thought-out mishaps are encouraged and rewarded. As well-known educational consultant and TED speaker Ken Robinson states, "If you are not prepared to be wrong, you will never come up with anything creative."[6] Part of creating a healthy risk culture is developing a well-thought-out and articulated risk tolerance. The risk tolerance of an organization is stating what level of risk the organization is willing to take, and also clearly stating what levels of risk are not acceptable. Obviously, anything that threatens the viability of the firm, or puts human health or safety at risk should not be tolerated. However, a culture that encourages managers to take prudent business risks within the level of risk tolerance will spur more creativity, more action, and likely a more enjoyable work environment.

---

[6] Robinson, K., "Do Schools Kill Creativity?," http://www.ted.com/talks/ken_robinson_says_schools_kill_creativity.

Finally, an organization needs to attempt to calculate the cost-effectiveness of its risk management. As mentioned previously, Hydro One successively did this by introducing the concept of *risk bang for the buck*, which was, in essence, a return on risk management expenditures.[7] By explicitly tracking costs, as well as the value added by risk management, a more efficient and healthy risk management function will be created. Implicitly assuming that risk management is a function that must be implemented to eliminate all downside risk leads to it becoming a bloated and overarching corporate bureaucracy with limited value, but overbearing reach.

Concluding Thoughts

Risk is an inherent part of business, and business management is inherently nothing more than risk management. However, in the current culture of fear and mandatory regulation, organizational risk management has often gone too far. In essence, the demands of regulators, corporate stakeholders, and even managers have created a culture that has created bubble-wrapped organizations. Excessive risk management is not benign. There are both significant explicit and implicit costs of having too strong of a risk management function. Furthermore, the phenomenon of risk homeostasis illustrates in example after example that an excessive focus on risk actually increases the possibility of bad things occurring—the exact opposite of the intended effect.

To be sure, risk management is a necessary part of managing a successful organization. However, good risk management is not excessive; instead it is smart and relevant for the situation. Tearing off the bubble wrap may involve a few more scrapes and the occasional broken bone, but in the long run, it makes the organization both healthier and ultimately safer and better risk-adjusted. If nothing else, it makes being a kid a lot more fun.

---

[7] Aabo, T., Fraser, J., Simkins, B., "The Rise and Transformation of the Chief Risk Officer: A Success Story on Enterprise Risk Management," Journal of Applied Corporate Finance, Winter 2005.

CHAPTER 10

What Is the Future of Risk Management?

The great Yankees catcher Yogi Berra famously said, "It's tough to make predictions, especially about the future." In this chapter, I am not so much going to make predictions, but present my own view of where we are trending and put forward some hopes that I have about the future of risk management.

As a field of study and as a management discipline, risk management is paradoxically one of the oldest disciplines and simultaneously still in its infancy. It is one of the oldest management disciplines, in that risk-taking has been acknowledged as core to organizational success since the earliest cavemen determined how best to gang hunt a woolly mammoth. It is also still in its infancy as the data-driven approach to risk management that is so prevalent has only recently become possible due to advances in computing techniques.

As an academic field, risk management has strong roots in engineering. In business academia, risk has traditionally been rooted in finance and operations, with little in the way of serious study in areas such as organizational behavior and marketing or strategy. The finance and operations components have almost completely ignored the sociology of risk management, preferring to focus on the quantitative studies, which are much easier to get concrete results for. As such, it has ignored the real issues surrounding risk culture and complexity.

As an illustration of the bias against risk management in academia, I was once asked to develop a university course in enterprise risk management (ERM). However, even before drafting an outline, the course was rejected as a regular course because ERM as a subject was considered to be nothing but "blah, blah, blah." I understand, and even appreciate the criticism, but simply because one does not (yet) have concrete answers for a question does not mean it is not worthy of study—perhaps, it makes it even more worthy of study! Of course, the most current trend is for advanced courses and programs in risk management to be very popular.

Trends in Risk Management

To begin examining the future, perhaps, it is useful to look at some current trends. The first trend is that of greater expectations (hopes) for what risk management can do. One need is to only examine the statements of politicians and regulators to see that there is an expectation that risk as an issue can be "solved." Risks, however, evolve, and Orgel's law teaches us that "evolution is smarter than we are."

The greater expectations set up risk management to be a field of disappointment. "Solving risk" is a term that does not make any sense whatsoever if one thinks about it. True, we can solve, or at least diagnose the complicated risks, such as an engine running out of fuel, but complex risks, such as how to create a winning product, or how to prevent or exploit market bubbles are by definition unsolvable. Additionally, risk and uncertainty are an inherent part of any organization. If risk and uncertainty are "solved," then will there even be a need for organizations?

Risk and risk tolerances are emotional in nature for the individual and sociological in nature at the organizational and societal levels. You cannot "solve" emotions or sociological issues. I do not even know how to define what such a solution would look like.

The trend, however, in popular belief and among those that do not know better is that risks are solvable. This, in part, leads to the bias and trend in thinking that risk is a bad thing. Without risk-takers, however, we would not have social media, technological developments, medical advances, or even the wheel and controllable fire for that matter. The scientists, engineers, organizers, entrepreneurs, and so on who have developed the modern conveniences that we take for granted were all risk-takers. They defied convention, and they certainly never took the path of lowest risk, or thought that risk was something to be avoided. Risk is not solvable, and it is those who get comfortable with that fact that we have to thank for our highly developed (and comfortable) lives.

Related to the trend of solving risk is the quantification of risk. Data and quantification are great things—for those risks that can be measured and that can be measured correctly. Unfortunately, many important risks cannot be measured, and many risks are inappropriately measured as discussed in Chapter 2. Despite my saying this, I believe that there are still many more risks that can be measured and should be measured. One area for growth in measurement is using Big Data techniques and social media to study the evolving trends and patterns. Data can give us clues to complexity, and that can be helpful, even though we cannot use that data directly to solve complexity. Social media data, however, can help us understand and build our intuition toward data, but will never replace the human who has their finger on the imperceptible pulses of society.

The trend is also for greater regulation of risk. Regulation definitely has its place, but I, and many others fear, that regulation is becoming too rigid. Rigidity in regulation prevents judgment to take its rightful place. Regulation also implicitly and explicitly assumes that risks are complicated. Regulation and complicated thinking has little place in the presence of complexity.

Regulation also builds rigidity into the economic system, which increases both the probability of a systemic risk occurring as well as the severity of a systemic risk, such as the 2008 financial crisis. Several years ago, my family moved into a new house on a heavily wooded lot. Shortly thereafter, a rare hurricane came through the area. I lost over a 100 trees on my property, as of course did my neighbors. The interesting thing is that the trees that got knocked down were the old trees with the deepest roots and the thickest trunks. The trees that were younger or not as solid had a much greater survival rate. The thick-trunk trees did not have the flexibility to adjust with the wind, while the less sturdy trees had the ability to bend, but not break. Despite their thick trunks and deep roots, the older, more established trees were no match for the wind. Regulation acts like a thick-trunked tree with deep roots. When a hurricane comes (and hurricanes will always come along), these are the trees that get uprooted and cause the most property damage.

These are some of the trends that I see in risk management. While it is a positive that risk management is now getting serious attention for its role in an organization's success, it is important to not get carried away. However, one of the things I have learned about trends in business is that just when you think it is a trend, you find out it is a cycle.

Some Potential Developments

One trend that is currently faddish is that of the trend of the entrepreneur. Business incubators, courses, and workshops on entrepreneurship are all the rage. It is almost as if we are revisiting the age at the turn of the 1900s when the railroad magnate and the industrialist were the masters of the universe.

Much of the current focus on entrepreneurship is on the millennial generation. Many of them (not all of them) bring a new approach to risk. Like entrepreneurs before them, they tend to be skeptical of accepted wisdom, and instead, they are intent on creating new ways of developing ideas and organizations. With this new spirit of entrepreneurship come new ideas about risk management. This can only be a good thing, as new ideas, even those that turn out to be wanting or incorrect, are necessary for developing and enhancing a discipline.

Many of these new entrepreneurial ventures are focused on exploiting the power of social media. This may seem to be unrelated to risk management, but I believe adaptation of social media techniques can dramatically improve risk management. By definition, social media is exploiting how societies and groups of individuals behave, interact, and adapt. Social media is complexity in action. Social media is giving us clues and ideas on how to harness or at least adapt to complexity. If nothing else, social media is exploiting complexity, and exploiting complexity, I believe, is a major area of potential development for risk management.

Social media is also demonstrating the importance of trying to understand social systems. In Chapter 4, the importance of "people, people, people" for risk management was highlighted. The role of people in risk management is key, and it is likely that risk management, catalyzed by social media entrepreneurs, will wake up to this fact. Perhaps, organizational risk management as a respected field of academic study will arise within sociology departments.

It is also quite possible that risk management as a discipline will go the other way. Instead of the focus on getting a risk management certification, there will be a realization that risk management is central with management. In other words, just like breathing is central to life, risk management is central to the success of an organization. You do not take courses in breathing (except maybe for some health conscious zealots), and likewise, risk management reverts to being simply management.

Despite my earlier comments about risk in academia, the rise of risk management certification programs and advanced courses in risk management does indeed demonstrate that there is a thirst for risk management knowledge. With that knowledge, however, comes a need to not forget about "thinking." Risk management is not necessarily about knowing things. Knowing about risk is complicated thinking. Computers are increasingly taking over the business of knowing things and managing complicated systems. Instead, it is incumbent on the modern risk management to understand things, rather than just know them. Creativity is more important in risk than knowledge. Being able to connect the dots is key. The ability to think about things is more important than the ability to know things.

One criticism I have about the current new wave of highly trained risk engineers is that they understand the mathematics, but they do not understand how businesses work and how markets and industries work. The more experienced risk managers may not understand the mathematics, and have as sophisticated toolbox, but they do have an intuitive feel for how organizations evolve, businesses evolve, industries evolve, and economies evolve. I fear that the emphasis on knowing about risk management will "crowd out" the understanding of risk management.

The emphasis on knowing things is also highly correlated with having a complicated mindset. However, I believe that having a complexity mindset is key. In our information age, knowledge is becoming a commodity. Any piece of knowledge can be found almost instantaneously using an Internet search engine. IBM's Watson computer can beat chess masters and of course win at the TV show *Jeopardy*. Increasingly, computers beat humans when it comes to all matters that are complicated. However, risk is rarely complicated; it is almost always complex, and it is in managing complexity that humans beat computers almost every time. Learning to acknowledge, understand, and become comfortable with dealing with complexity will give the manager of the future, and especially the risk manager of the future, a key competitive advantage.

The final expectation I have for risk management is that risk management will become a part of every manager's path to success. Traditionally, in order to become a senior executive, or even the CEO of an organization, one had to show competency in marketing, finance, and in operations. I expect, and hope, that risk management will be added to that list of must-have competencies for one to be considered suitable for leadership.

Concluding Thoughts

In this book, I have intentionally stayed away from many of the typical specifics about managing risks and instead stayed at a high-level conceptual view. That approach was very intentional. Risk management techniques and tactics are specific to each individual field and to each individual organization. Hopefully, however, you, the reader, have come away with a few high-level conceptual ideas about managing risk that you can apply to your specific situation and issues. Hopefully, you will be inspired to ask your own questions about conventional risk wisdom and will attempt to try new ideas and be willing to take new risks when it comes to designing and implementing risk management. Hopefully, you have advanced your own ideas about the future of risk management.

I began by stating in the introduction that "This book is based on three fundamental tenets: (1) that risk management is a vital task for developing competitive advantage, (2) having knowledge, but more importantly skill and intuition in risk management is key for advancing one's career, particularly in light on the onslaught of the robots and computers that are replacing both blue-collar and white-collar jobs, and (3) there is a need to take a fresh look at risk management by questioning some old assumed axioms and asking some fresh questions." I hope that these themes have come across even though I did not explicitly dwell on them.

My wish is that in some small way that this book has added a grain of sand to the mountain of knowledge on risk and risk management. Risk management is a fascinating field. It is fun, it is important, and it is exciting to think of how the field will continue to develop and evolve.

Finally, I would like to thank you the reader for taking the time to read this book. Again, repeating myself from the introduction, I hope you had as much fun reading this book as I had writing it.

Index

OTHER TITLES IN OUR FINANCE AND FINANCIAL MANAGEMENT COLLECTION

John A. Doukas, Old Dominion University, Editor

- *Essentials of Retirement Planning: A Holistic Review of Personal Retirement Planning Issues and Employer-Sponsored Plans, Third Edition* by Eric J. Robbins
- *Financial Services Sales Handbook: A Professionals Guide to Becoming a Top Producer* by Clifton T. Warren
- *Money Laundering and Terrorist Financing Activities: A Primer on Avoidance Management for Money Managers* by Milan Frankl and Ayse Ebru Kurcer
- *Introduction to Foreign Exchange Rates, Second Edition* by Thomas J. O'Brien
- *Rays of Research on Real Estate Development* by Jaime Luque
- *Weathering the Storm: The Financial Crisis and the EU Response, Volume I: Background and Origins of the Crisis* by Javier Villar Burke
- *Weathering the Storm: The Financial Crisis and the EU Response, Volume II: The Response to the Crisis* by Javier Villar Burke

Announcing the Business Expert Press Digital Library

Concise e-books business students need for classroom and research

This book can also be purchased in an e-book collection by your library as

- a one-time purchase,
- that is owned forever,
- allows for simultaneous readers,
- has no restrictions on printing, and
- can be downloaded as PDFs from within the library community.

Our digital library collections are a great solution to beat the rising cost of textbooks. E-books can be loaded into their course management systems or onto students' e-book readers.
The **Business Expert Press** digital libraries are very affordable, with no obligation to buy in future years. For more information, please visit **www.businessexpertpress.com/librarians**. To set up a trial in the United States, please email **sales@businessexpertpress.com**.

CPSIA information can be obtained
at www.ICGtesting.com
Printed in the USA
BVHW092126081218
535118BV00009B/338/P

9 781631 575419